Numerical Studies on Projectiles: A Case Study of Bullet Dynamics

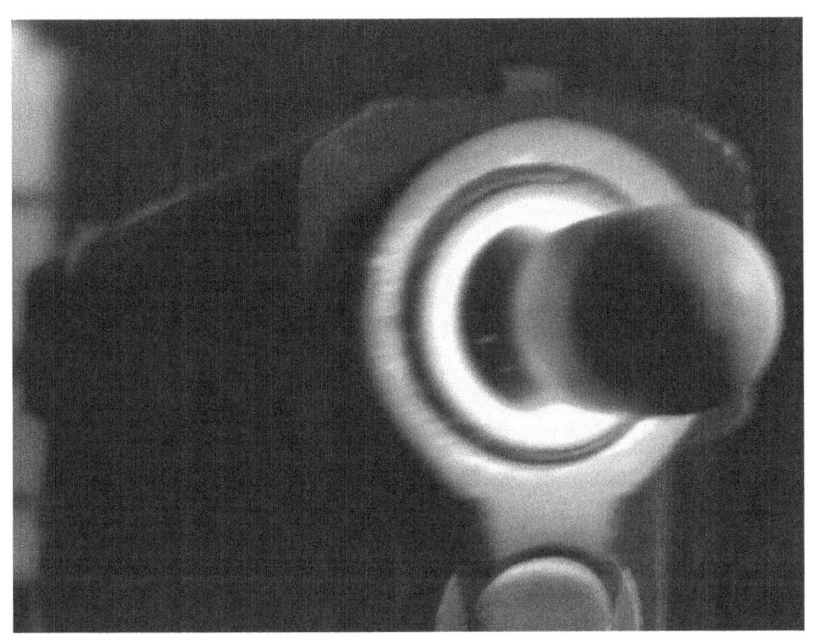

By

Dr Gurunadh Velidi

Dr Ugur Guven

Disclaimer

Although the author and publisher have made every effort to ensure that the information in this book was correct at press time, the author and publisher do not assume and hereby disclaim any liability to any party for any loss, damage, or disruption caused by errors or omissions, whether such errors or omissions result from negligence, accident, or any other cause. For this book, student analyses have been included as the main data.

Copyright

CFD Analysis of Projectiles: Case Study of a Bullet
© May 2023, by Dr Gurunadh Velidi and Dr. Ugur GUVEN
Amazon Publications

ASIN: B083LCV2D1

ISBN: 9798336146851

Miami-Paris-London

All rights reserved.
No part of this publication may be reproduced, stored in a retrieval system, stored in a database and / or published in any form or by any means, electronic, mechanical, photocopying, recording or otherwise, without the prior written permission of the publisher.

Foreword

In the realm of engineering and physics, Computational Fluid Dynamics (CFD) has emerged as a powerful tool for analyzing and understanding the complex behaviors of fluid flow. Its applications are far-reaching, spanning a multitude of industries and disciplines. Among its many practical applications, the analysis of projectiles, particularly bullets, holds a special significance due to its critical role in military operations, law enforcement, and even recreational shooting.

The book you hold in your hands, "CFD Analysis of Projectiles: Case Study of a Bullet," delves deep into the fascinating world of CFD analysis and its specific application to bullet design and performance. With its comprehensive exploration of the subject matter, this book serves as a valuable resource for researchers, engineers, and enthusiasts who seek a thorough understanding of the aerodynamics and fluid flow characteristics associated with projectiles.

Authored by a team of seasoned experts, this book provides a meticulous examination of the principles and methodologies employed in CFD analysis. It begins by laying the groundwork, introducing the fundamental concepts of fluid dynamics and their relevance to projectile motion. The subsequent chapters delve into the intricacies of bullet design, the effects of various factors such as shape, size, and mass distribution on aerodynamic performance, and the challenges faced in modeling the complex flow phenomena involved.

One of the key strengths of this book lies in its emphasis on practicality and real-world applicability. The authors employ a case study approach, focusing on a specific bullet design and using it as a vehicle to explore the intricacies of CFD analysis. By presenting a concrete example, they enable readers to grasp the underlying principles and techniques, making the material more accessible and engaging. The case study follows a step-by-step methodology, starting with the creation of a virtual model and progressing through the simulation setup, solver selection, and post-processing of results. This structured approach allows readers to follow along and gain a comprehensive understanding of the entire process.

Furthermore, the book delves into the various challenges encountered in CFD analysis, such as turbulence modeling, mesh generation, and boundary conditions. It provides valuable insights into the best practices and techniques for overcoming these challenges, equipping readers with the tools they need to tackle similar problems in their own work.

While the primary focus of this book is on CFD analysis, it also sheds light on the symbiotic relationship between experimental testing and computational simulations. The authors emphasize the importance of validating CFD results against real-world data, demonstrating the significance of complementary approaches for accurate analysis and prediction. This comprehensive approach further strengthens the book's relevance, bridging the gap between theory and practice.

As the world continues to advance technologically, the study and analysis of projectiles through CFD techniques will remain a vital field of research and development. The insights gained from this book have the potential to shape the future of bullet design, leading to enhanced accuracy, range, and terminal performance. Moreover, the knowledge and methodologies presented here can be extended to other areas that involve fluid flow and aerodynamics, broadening the scope of its applicability.

In conclusion, "CFD Analysis of Projectiles: Case Study of a Bullet" stands as a comprehensive and invaluable resource for anyone seeking to deepen their understanding of CFD analysis, specifically in the context of bullet design and performance. With its meticulous approach, practical insights, and real-world case study, this book sets a high standard for excellence in the field. It is my sincere hope that readers find this book as enlightening and inspiring as I have, and that it sparks new avenues of research and innovation in the fascinating realm of projectile aerodynamics.

TABLE OF CONTENTS

Page

TABLE OF CONTENTS i
LIST OF FIGURES v
LIST OF GRAPHS..ix
LIST OF TABLE..x

1. Chapter 1

    1.1 Introduction 1

    1.2 History of Bullet...1

    1.3 Basic type of bullet and cartridge with their terminology..........................2

        1.3.1   Anniversary Round...2

        1.3.2   Anvil..2

        1.3.3   Assult Rifle Cartridge...3

        1.3.4   Boat Tailed bullet...3

        1.3.5   Bore..3

        1.3.6   Bottlenecked Case..4

        1.3.7   Bronze Point Bullet..4

        1.3.8   Coal-Dust Dummy...4

        1.3.9   Combustible Cartridge..5

        1.3.10 Copper-tubed Bullet...5

        1.3.11 Corrugated Case..5

        1.3.12 Crispin Cartridge...6

        1.3.13 Cupfire cartridge...6

        1.3.14 Extractor Cord Cartridge..7

1.3.15 Flobert...7

1.3.16 Frangible Bullet... ..7

1.3.17 Gravity Feed Cartridge..8

1.3.18 Gyrojet Cartridge...8

1.3.19 Heel Type Bullet..8

1.3.20 Hollow-Point Bullet..9

1.3.21 Hoxie Bullet..9

1.3.22 Inside Primed Cartridge..9

1.3.23 Inside-Lubricated Bullet...10

1.3.24 Lipfire Cartridges...10

1.3.25 Mule Ear Cartridge...10

1.3.26 Ogive..11

1.3.27 Paper Patched Bullet..11

1.3.28 Rimfire...11

1.3.29 Seprated-Primed...12

1.3.30 Sintered Bullet...12

1.3.31 Soft Nosed Bullet...13

1.3.32 Spitzer Bullet...13

1.3.33 Volcanic Cartridge...13

1.4 Forces and External Factors Acting on The Bullet 14

1.5 External Factors..14

   1.5.1 Wind...14

   1.5.2 Vertical Angles...14

   1.5.3 Ambient Air Density..14

1.6 Long Range Factors...15

    1.6.1 Gyroscopic Drift………………………………………………………..…15

    1.6.2 Magnus Effect………………………………………………………...…15

    1.6.3 Poisson Effect……………………………………..…………………..…15

    1.6.4 Coriolis Drift…………………………………………………………..…15

..

2. Chapter-2

    2.1 Equation of Motion of Bullet………………………………………………...........16

    2.2 Orientation and position of Bullet……………………………………………...…21

3. Chapter 3

    3.1 Overview of M16 Rifle……………………………………………………….....…25

    3.2 Dimension of M16 Cartridge…………………………………………………...…..26

    3.3 Types of M16 bullet……………………………………………………..……...….27

4. Chapter 4

    4.1 Boundary layer………………………………………………………….…….…….29

    4.2 Separation of boundary layer……………………………………………….…..….31

    4.3 Formation of shock waves………………………………………………….………32

    4.4 Change in properties across the Shock...………………………………….………34

5. Chapter 5

    5.1 Analysis of shapes……………………………………………………………35

        5.1.1 Point Head Bullet……………………………………………………….35

        5.1.2 For pure cylindrical body……………………………………………….40

6. Chapter 6

    6.1 Analysis of Semi blunt head bullet……………………………………………44

        6.1.1 Mach No 3.5……………………………………………………………..44

            6.1.1. a 0 degree angle of attack……………………………………………44

            6.1.1. b 2 degree angle of attack………………………………. ………...47

            6.1.1. c 2.5 degree angle of attack…………………………………..……...50

        6.1.2 Mach no 3…………………………………………………………….…52

            6.1.2. a 0 degree angle of attack……………………………………………..52

            6.1.2. b 2 degree angle of attack……………………………………………..54

7. Conclusion…………………………………………………………………...……57

8. Appendix……………………………………………………………………………60

9. Reference……………………………………………………………………………72

List Of Figure

Figure.No.1.1 Lead Round Bullet Ammunation..3

Figure.No.1.2 Minie Ball..3

Figure.No.1.3 Some Modern Bullet...3

Figure.No.1.4 Anniversary Round...3

Figure.No.1.5 Anvil..3

Figure.No.1.6 Assult Rifle Cartridge..3

Figure.No.1.7 Boat Tailed bullet..3

Figure.No.1.8 Bore...4

Figure.No.1.9 Bottlenecked Case...4

Figure.No.1.10 Bronze Point Bullet...4

Figure.No.1.11 Coal-Dust Dummy..5

Figure.No.1.12 Combustible Cartridge..5

Figure.No.1.13 Copper-tubed Bullet...5

Figure.No.1.14 Corrugated Case...6

Figure.No.1.15 Crispin Cartridge..6

Figure.No.1.16 Cupfire cartridge...6

Figure.No.1.17 Extractor Cord Cartridge..7

Figure.No.1.18 Flobert..7

Figure.No.1.19 Frangible Bullet..7

Figure.No.1.20 Gravity-feed Cartridge..8

Figure.No.1.21 Gyrojet cartridge...3

Figure.No.1.22 Heel Type Bullet..8

Figure.No.1.23 Hollow-Point Bullet...9

Figure.No.1.24 Hoxie Bullet..9

Figure.No.1.25 Inside Primed Cartridge..9

Figure.No.1.26 Inside-Lubricated Bullet..10

Figure.No.1.27 Lipfire Cartridges...10

Figure.No.1.28 Mule Ear Cartridge..10

Figure.No.1.29 Ogive..11

Figure.No.1.30 Paper Patched Bullet..11

Figure.No.1.31 Rimfire..11

Figure.No.1.32 Seprated-Primed..12

Figure.No.1.33 Sintered Bullet...12

Figure.No.1.34 Soft Nosed Bullet...12

Figure.No.1.35 Spitzer Bullet...12

Figure.No.1.36 Volcanic Cartridge..3

Figure.No.2.1 Bullet with inertial frame and fixed frame..3

Figure.No.2.2 Rotation/ Euler Angles...3

Figure.No.3.1 M16 Rifle...12

Figure.No.3.2 M16 Cartridge..12

Figure.No.3.3 Type of M16 Cartridge..12

Figure.No.4.1 Boundary Layer..12

Figure.No.4.2 Turbulent/Laminar Flow...12

Figure.No.4.3 Boundary Layer Over a Ogive Body 12

Figure.No.4.4 Boundary Layer Separation...12

Figure.No.4.5 Bow Shock...12

Figure.No.4.6 Transonic Flow/ lamda shock..12

Figure.No.4.7 Transonic Flow/ Bow Shock..12

Figure.No.4.8 Supersonic Flow/ Oblique Shock..12

Figure.No.4.9 Oblique Shock Wave...12

Figure.No.4.10 Normal Shock Wave..12

Figure.No.5.1 Catia Model of Bullet..12

Figure.No.5.2 MeasureInertia of Bullet in Catia...12

Figure.No.5.3 Grid of a Point Head Bullet..12

Figure.No.5.4 Contours of Dynamic Pressure of a Point Head Bullet ....,......12

Figure.No.5.5 Contours of Static Pressure of a Point Head Bullet..................................12

Figure.No.5.6 Displacement Thickness 12

Figure.No.5.7 Grid of a Pure Blunt Head Bullet...12

Figure.No.5.8 Contours of Static Pressure of a Pure Blunt Head Bullet..........................12

Figure.No.5.9 Contours of Dynamic Pressure of a Pure Blunt Head Bullet....................12

Figure.No.5.10 Pathlines of Pure Blunt Head Bullet..12

Figure.No.6.1 Contour of Static Pressure of Semi Blunt Head Bullet at 3.5 Mach & at
          Zero degree angle of attak...12

Figure.No.6.2 Contour of Dynamic Pressure of Semi Blunt Head Bullet at 3.5 Mach & at
          Zero degree angle of attak...12

Figure.No.6.3 Contour of Static Pressure of Semi Blunt Head Bullet at 3.5 Mach & at 2
          degree angle of attak...12

Figure.No.6.4 Contour of Dynamic Pressure of Semi Blunt Head Bullet at 3.5 Mach & at
          2 degree angle of attak..12

Figure.No.6.5 Contour of Static Pressure of Semi Blunt Head Bullet at 3.5 Mach & at
          2.5 degree angle of attak...12

Figure.No.6.6 Contour of Dynamic Pressure of Semi Blunt Head Bullet at 3.5 Mach & at
          2.5 degree angle of attak...12

Figure.No.6.7 Contour of Static Pressure of Semi Blunt Head Bullet at 3 Mach & at Zero
          degree angle of attak..12

Figure.No.6.8 Contour of Dynamic Pressure of Semi Blunt Head Bullet at 3 Mach & at Zero degree angle of attak……………..………………………………..12

Figure.No.6.9 Contour of Static Pressure of Semi Blunt Head Bullet at 3 Mach & at 2 degree angle of attak……………….………………………………………12

Figure.No.6.10 Contour of Dynamic Pressure of Semi Blunt Head Bullet at 3 Mach & at 2 degree angle of attak…………………………………………………..12

List of Graphs

Graph.No.6.1 Contour of Static Pressure of Semi Blunt Head Bullet at 3.5 Mach & at Zero degree angle of attak……………………………………………12

Graph.No.6.2 Contour of Dynamic Pressure of Semi Blunt Head Bullet at 3.5 Mach & at Zero degree angle of attak……………………………………………..12

Graph.No.6.3 Contour of Static Pressure of Semi Blunt Head Bullet at 3.5 Mach & at 2 degree angle of attak……………………………………………12

Graph.No.6.4 Contour of Dynamic Pressure of Semi Blunt Head Bullet at 3.5 Mach & at 2 degree angle of attak……………………………………………..12

Graph.No.6.5 Contour of Static Pressure of Semi Blunt Head Bullet at 3.5 Mach & at 2.5 degree angle of attak……………………………………………12

Graph.No.6.6 Contour of Dynamic Pressure of Semi Blunt Head Bullet at 3.5 Mach & at 2.5 degree angle of attak……………………………………………..12

Graph.No.6.7 Contour of Static Pressure of Semi Blunt Head Bullet at 3 Mach & at Zero degree angle of attak……………………………………………12

Graph.No.6.8 Contour of Dynamic Pressure of Semi Blunt Head Bullet at 3 Mach & at Zero degree angle of attak……………………………………………..12

Graph.No.6.9 Contour of Static Pressure of Semi Blunt Head Bullet at 3 Mach & at 2 degree angle of attak……………………………………………12

Graph.No.6.10 Contour of Dynamic Pressure of Semi Blunt Head Bullet at 3 Mach & at 2 degree angle of attak……………………………………………..12

Graph.No A Drop of M16 Projectile "Pre Measured"…………………………………..12

List of Table

        Table.No.1 Overview of different types of M16 rifles..............................12

CHAPTER 1

Introduction

The field of Computational Fluid Dynamics (CFD) has revolutionized engineering and scientific research by providing powerful tools for analyzing and understanding fluid flow phenomena. While its applications span a wide range of industries, the analysis of bullets through CFD holds a special significance due to its crucial role in fields such as military operations, law enforcement, and shooting sports. In this essay, we will explore the importance of CFD analysis of bullets and the benefits it offers to various stakeholders.

First and foremost, CFD analysis of bullets enables engineers and designers to optimize bullet performance and enhance accuracy. The aerodynamic properties of a bullet significantly impact its trajectory, stability, and terminal ballistics. By utilizing CFD simulations, researchers can investigate and fine-tune the design parameters such as shape, nose profile, and base configuration to minimize drag, maximize stability, and improve overall flight characteristics. This optimization process can lead to bullets that exhibit enhanced long-range accuracy, reduced wind drift, and improved target penetration, all of which are critical factors in military engagements, law enforcement scenarios, and competitive shooting events.

Furthermore, CFD analysis provides valuable insights into the fluid flow phenomena around bullets, which can contribute to reducing supersonic crack noise, recoil, and muzzle blast. Understanding the complex interactions between the bullet and the surrounding air can help engineers develop designs that minimize these undesirable effects, thereby improving shooter comfort, reducing noise pollution, and optimizing firearm performance.

Moreover, CFD analysis plays a pivotal role in studying the impact of environmental factors on bullet behavior. By considering variables such as altitude, temperature, humidity, and wind conditions, researchers can accurately predict how these factors influence the flight dynamics of bullets. This information is invaluable for military snipers who must make precise long-range shots in various operational environments, where factors like air density and wind can have a significant impact on bullet trajectory. Similarly, law enforcement officers and competitive shooters can benefit from this knowledge to adapt their shooting techniques and equipment for different conditions, enhancing their effectiveness and accuracy.

Another crucial aspect where CFD analysis of bullets proves important is in the evaluation and development of bullet designs for specific applications. For example, armor-piercing rounds require a unique understanding of the fluid flow and impact dynamics involved in penetrating armored materials. Through CFD simulations, engineers can analyze the behavior of these bullets upon impact, study deformation patterns, and optimize design features to enhance penetration capabilities.

Furthermore, CFD analysis allows for cost-effective and efficient exploration of a wide range of design alternatives. Traditional experimental testing can be time-consuming and expensive, especially when dealing with multiple design iterations. CFD simulations offer a virtual testing environment where engineers can evaluate different design options rapidly and at a fraction of the cost. This enables the exploration of a larger design space, facilitating innovation and advancement in bullet technology.

In conclusion, the importance of CFD analysis of bullets cannot be overstated. It offers engineers, designers, and researchers a powerful set of tools to optimize bullet performance, enhance accuracy, reduce undesirable effects, account for environmental factors, and develop specialized designs for specific applications. The knowledge gained from CFD analysis contributes to advancements in military engagements, law enforcement operations, and shooting sports, ultimately leading to more effective and efficient bullet designs. As the field of CFD continues to evolve, its application to bullet analysis will undoubtedly play a crucial role in shaping the future of firearm technology.

In addition, it must be stated that the utilization of a case study of bullets will also help us understand the aerodynamic behavior of projectiles as well, which can be very important for defense technologies. Hence, the analyses below for bullets also sheds light on the behavior of projectiles as well giving us a broad understanding of the aerodynamic principles behind their operability.

1.1 What is a Bullet?:

Bullet: A bullet is an object which is made by a metallic component projectile thrown or fired by a firearm like gun. Bullet normally do not contain any explosives but it can damage due to its penetration and impact.

1.2 History of Bullets:

Bullets were invented in 1500's and in starting round lead balls were to be used. Later in 1823's pointed bullet came into more use by a British army person Captain John Norton. This bullet basically had a hallow base which get expand under pressure to engage with a barrel's rifling but later on it got rejected by the British army due to round bullets which were used for the last 300 years.

Fig 1.1: Lead Round bullets Ammunition adapted from Url-1

Fig 1.2: Minie Ball adapted from Url-1

As the time passed, it was observed that in 1888, the pointed bullet came in much more use which were innovated by some of the gunsmith and it was also adopted by British army.

Fig 1.3: Some Modern Bullet adapted from Url-1

1.3 Basic shape of bullets and cartridges with their terminology

There are some common words which are required in this basic terminology. There are basic terms which are used to define cartridge lore are as follows:

1.3.1 ANNIVERSARY ROUND: It is a special type of commemorative cartridge which have a stamp of (US FA) issued to mark a particular event.

Fig 1.4: Anniversary Round adapted from Url-2

1.3.2 ANVIL: In a primer pocket there is a particular portion or we can define it as the resistance for the crushing action of firing pin which is in the firearm is provided by the primer only which causes the printing mixture to explode violently.

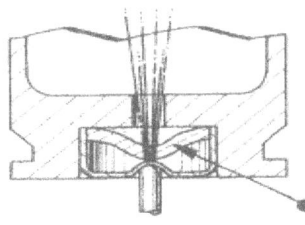

Fig 1.5: Anvil adapted from Url-2

1.3.3 ASSULT RIFLE CARTRIDGE: This cartridge is usually of short case length and of small calibre and its designed mainly for moderate firing rate.

Fig 1.6: Assult Rifle Cartridge adapted from Url-2

1.3.4 BOAT TAILED BULLET: Ballistic drag is usually reduced by this type of bullet only. The base of this bullet is tapered and is very similar to boat's stern profile.

Fig 1.7: Boat Tailed Bullet adapted from Url-2

1.3.5 BORE: Bore is basically a diameter of a particular weapon.

Fig 1.8: Bore adapted from Url-2

1.3.6 BOTTLENECKED CASE: In this cartridge there is a abrupt reduction in its diameter at its open end or at its mouth end.

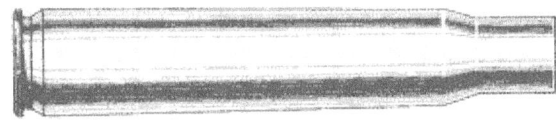

Fig 1.9: Bottlenecked Case adapted from Url-2

1.3.7 BRONZE POINT BULLET: Aerodynamic streamlining are provided by this kind of aerodynamic bullet and it also initiate expansion on force which separate bronze wedge which is located at the point of the bullet .

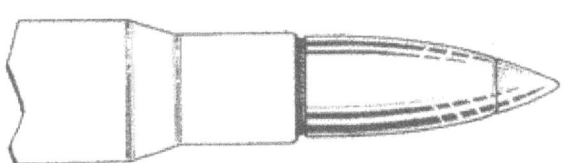

Fig 1.10: Bronze Point Bullet adapted from Url-2

1.3.8 COAL-DUST DUMMY- This bullet has a label reading of affixed paper "COAL-DUST" on it and it is manufactured in Britain. In this bullet instead of gun powder we have acting coal dust to approximate the load and to give a feel of a loaded round.

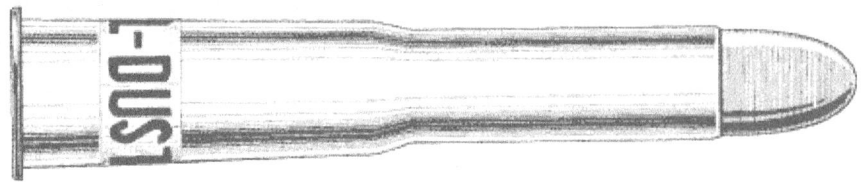

Fig 1.11: Coal-Dust Dummy adapted from Url-2

1.3.9 COMBUSTIBLE CARTRIDGE: When this type of bullet is fired then the entire envelope is to be consumed. This cartridge having a combustible mixture the term applied to certain ammunition which normally contained propellant sometime nitrated paper casting which enclosed the projectile.

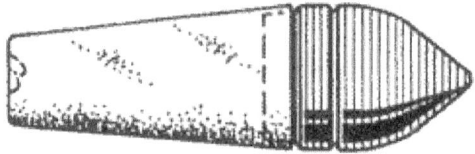

Fig 1.12: Combustible Cartridge adapted from Url-2

1.3.10 COPPER-TUBED BULLET: Aerodynamic shape is preserved in this type of bullet. It is also called as a express bullet as we can create an explosive bullet out of it with the help of fulminate. A thin closed ended copper tube which is inserted into a lead hollow pointed bullet is called a copper- tube bullet.

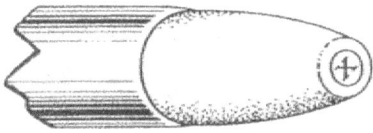

Fig 1.13: Copper-Tubed Bullet adapted from Url-2

1.3.11 CORRUGATED CASE: In dummy round this kind of bullets are used. These bullets have longitudinal grooves in it which mainly provide tactical and visual identification to the bullet

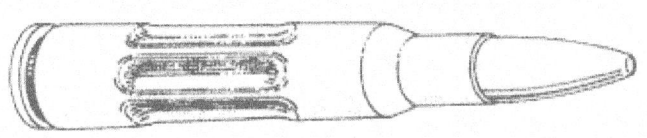

Fig 1.14: Corrugated Case adapted from Url-2

1.3.12 CRISPIN CARTRIDGE: In this bullet we have a annular ring which is located nearly in the midway between the base and that of the cartridge. This bullet have fulminate for more explosive a distinctive type of rim fire cartridge.

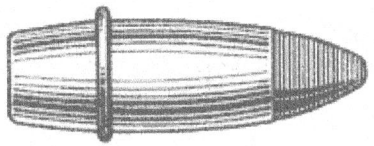

Fig 1.15: Crispin Cartridge adapted from Url-2

1.3.13 CUPFIRE CARTRIDGE: In earlier pistol this type of cartridge was used in which we have concave base in front loading rimfire so as to circumvent the rolling of the white board through a cylinder design. This cartridge is also known as 0.28, 0.42 and 0.30 caliber.

Fig 1.16: Cupfire Cartridge adapted from Url-2

1.3.14 EXTRACTOR CORD CARTRIDGE: This cartridge has a short length of a chord attached to it so that it can be assist in extracting a fired case and by figure it can be easily understand.

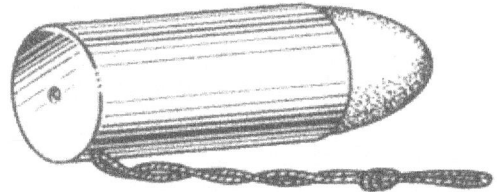

Fig 1.17: Extractor Cord Cartridge adapted from Url-2

1.3.15 FLOBERT: It is a small calibre rim fire cartridge which is widely used in Europe especially for indoor shooting.

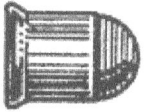

Fig 1.18: Flobert adapted from Url-2

1.3.16 FRANGIBLE BULLET: This bullet is mainly designed for load breakup rather than penetrate and it is made by different kind of powdered and compressed materials.

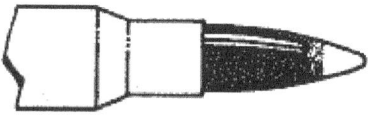

Fig 1.19: Frangible Bullet adapted from Url-2

1.3.17 GRAVITY-FEED CARTRIDGE: It is a lead and self-contained cartridge which is mainly used by loran, gaupillat and other gravity feed weapons.

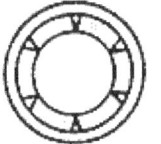 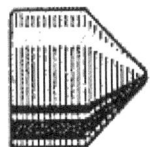

Fig 1.20:Gravity –feed Cartridge adapted from Url-2

1.3.18 GYROJET CARTRIDGE: It is manufactured by a firm of San Ramon California MB associate. At its centre it has some kind of a propellant which slightly angled discharge orifice. It is also called "Rocket projectile". Once it is fired the angled projectile to rotate and accelerate.

Fig 1.21: Gyrojet cartridge adapted from Url-2

1.3.19 HEEL-TYPE BULLET- Its front portion get flush with the case when loaded at the time and it also have a rear section of reduced diameter.

Fig 1.22: Heel-Type Bullet adapted from Url-2

1.3.20 HOLLOW-POINT BULLET- This type of bullet is more widely used to assist the shocking power and expansion when get in contact with flesh due to it recessed open cavity at its tip.

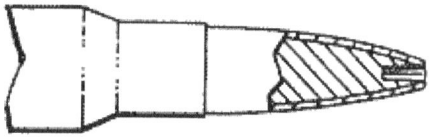

Fig 1.23: Hollow-Point Bullet adapted from Url-2

1.3.21 HOXIE BULLET: In this bullet at the tip some steel ball is embedded to some expansion when it comes in contact. This bullet is widely used for centre fire rifle calibre.

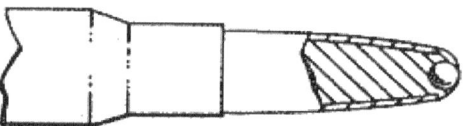

Fig 1.24: Hoxie Bullet adapted from Url-2

1.3.22 INSIDE-PRIMED CARTRIDGE: In this bullet the primer is less integral with the case. These generic term relates to a very big family of non-reloadable bullets and this act of the firing deformes the case to cause ignition.

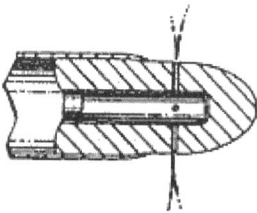

Fig 1.25: Inside-Lubricated Bullet adapted from Url-2

1.3.23 INSIDE-LUBRICATED BULLET: It is a patented bullet which is manufactured by UMC and this bullet is hollow and is filled with grease. A piston is there which forces the grease through some tiny holes in a side of bullet.

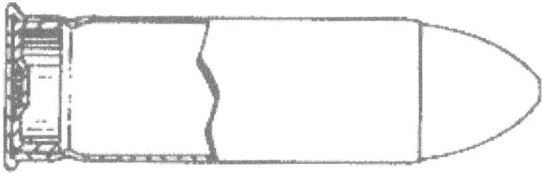

Fig 1.26: Inside-Primed Cartridge adapted from Url-2

1.3.24 LIPFIRE CARTRIDGE- It is a form of rim fire cartridge which is filled with lip project from the base rather than a full circumferential rim.

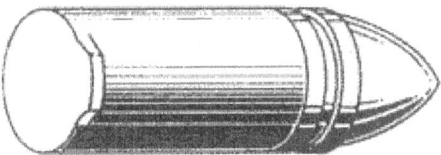

Fig 1.27: Lipfire Cartridge adapted from Url-2

1.3.25 MULE EAR CARTRIDGE: It is more suitable for 0.36- calibre for sharp dropping block pistol. This are generally used for rare shooting and the term is applied for the calibre.

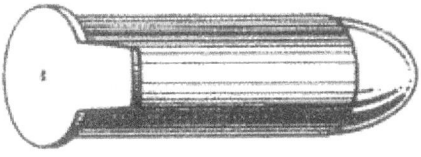

Fig 1.28: Mule Ear Cartridge adapted from Url-2

1.3.26 OGIVE- This is basically called the curved forward part of the bullet.

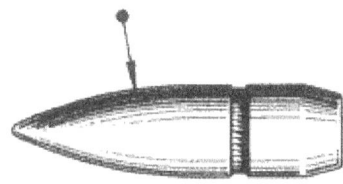

Fig 1.29: Ogive adapted from Url-2

1.3.27 PAPER PATCHED BULLET: In paper patched bullet we have a wrapped paper on the base and a bearing surface too. The paper is basically used to engage the bore rifling and one more strip off as the bullet exit.

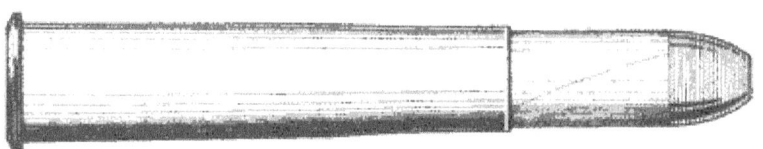

Fig 1.30: Paper Patched Bullet adapted from Url-2

1.3.28 RIMFIRE: In this bullet we have a priming compound at the base of the bullet and it is a major class of metal cased cartridges

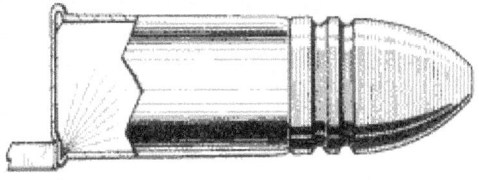

Fig 1.31: Rimfire adapted from Url-2

1.3.29 SEPRATED-PRIMED: In earlier cartridge bullet and propellant were within the bullet case, but in this bullet, we have percussion cap on a nipple with the flame which is conducted to the base of the bullet by the internal channeling.

Fig 1.32: Separated-Primed adapted from Url-2

1.3.30 SINTERED BULLET- It is a solid type of bullet which is formed when a high pressure is consolidation on powdered metal to get a bullet shape.

Fig 1.33: Sintered Bullet adapted from Url-2

1.3.31 SOFT NOSED BULLET- By the name it suggests that it is a jacketed bullet which is more used in hunting and by it figure we can see that it has a exposed lead tip whose main purpose is to expand upon impact with flesh.

Fig 1.34: Soft Nosed Bullet adapted from Url-2

1.3.32 SPITZER BULLET- The word spitz is a German word which means sharp and it is one of the modern bullet types.

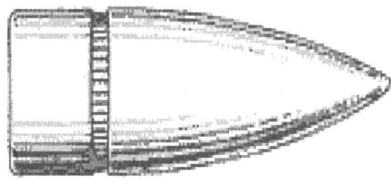

Fig 1.35: Spitzer Bullet adapted from Url-2

1.3.33 VOLCANIC CARTRIDGE: It is an evolution of hunt rocket ball. Within its base it has powder and primer and it also have concave shape at its base.

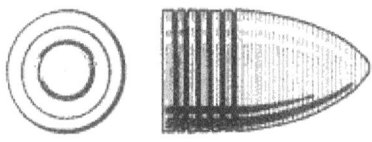

Fig 1.36: Volcanic Cartridge adapted from Url-2

1.4 Forces and external factors acting on the bullet

In a flight, the main force which are acting on a bullet are gravity, drag and if present also wind. Gravity basically imparts a downward acceleration on the projectile and causes it to drop from line of sight. Drag or air resistance decelerates the bullet with a force proportional to the square of the velocity and also cause deviation from its trajectory path.

1.5 External Factors

1.5.1 Wind: Wind has a wide range of effects on a bullet. It can deviate the bullet from its line of sight and can causes wind drift called Drag. Drag makes the bullet turn into the wind, and the centre of pressure come on its nose which causes a downward deflection. The direction of wind also deflects the trajectory of bullet. A headwind will slightly increase the relative velocity of the bullet, and increase drag and the corresponding drag. Similarly tail wind reduces the drag and the bullet drop.

1.5.2 Vertical Angles: The vertical angle of shot of a bullet will also affect the trajectory of the shot. Gravity on a bullet acts perpendicular to the bullet path. If the angle is up or down, then the perpendicular acceleration will actually be less. The effect of path wise acceleration component will be very small, so shooting up or downhill will both results in similar decrease in bullet drop.

1.5.3 Ambient air density: Air temperature, pressure and humidity variations make up the ambient air density. Humidity has a counter intuitive impact. As water vapor has a density of 0.08 gram per litre, while dry air averages about 1.225 grams per liter, higher humidity actually decreases the air density and therefore drag decreases.

1.6 Long range factors

1.6.1 Gyroscopic drift (Spin Drift): Gyroscopic drift is a spin induced drift experienced by a bullet. A spin stabilized bullet acted by a spin induced sideways component. When bullet rotates in a clockwise direction, it experiences right side component of drift and vice a versa. This is because the bullet's longitudinal axis and the direction of the velocity of center of gravity deviate by a small angle, which is called equilibrium yaw or yaw of repose.

1.6.2 Magnus effect: Magnus effect affected the spin stabilized bullet as the spin of bullet creates a force acting either up or down which is perpendicular to the sideways vector of the wind. The Magnus effect induced pressure differences around the bullet cause a downward force when wind is from right and cause a upward force when wind is from left.

1.6.3 Poisson effect: Drift also occurs due to Poisson effect which depends on the nose of the projectile being above the trajectory. The up tilted nose of the bullet causes an air cushion to build up underneath the bullet and due to that there is increase in friction between this cushion and the bullet, therefore the spin of bullet tend to roll the cushion and move side sideways. Both Poisson and Magnus effect will reverse their direction of drift if the nose fails below the trajectory.

1.6.4 Coriolis drift: Coriolis drift is caused by the Coriolis and Eötvös effect. These effect cause drift related to the spin of earth. It can be up, down, left or right. It is not an aerodynamic effect.

CHAPTER 2

2.1 Equation of Motion of Bullet

With the reference to the equation of motion and its dynamic nature, it is possible to derive the equation of motion for bullet. Assuming the weight of bullet is negligible, and no drift (other external deviating forces) are acting, then we can state as below:

Considering the fixed frame and body frame for a round head bullet.

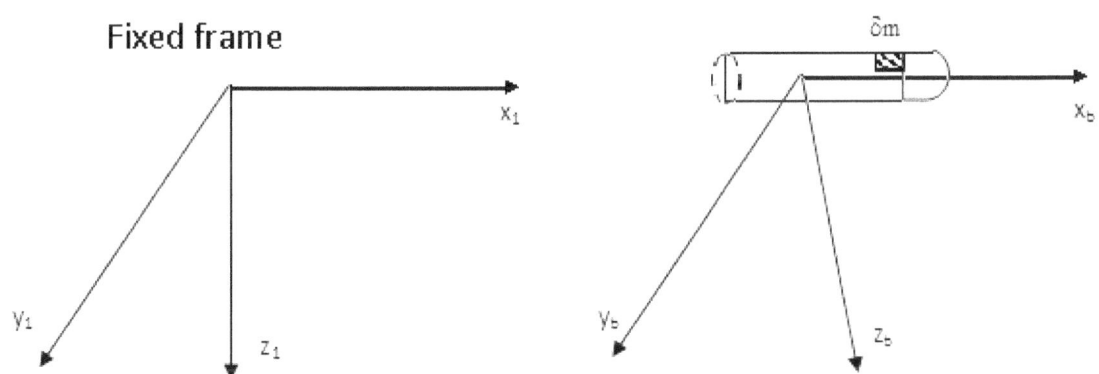

Figure no 2.1: Bullet with inertial frame and fixed frame

From Newton's second law of motion

$$\sum f = \frac{d}{dt}(mv)$$

(2.1)

$$\sum M = \frac{d}{dt}(H)$$

(2.2)

In Scalar

$$F_x = \frac{d}{dt}(mu) \tag{2.3}$$

$$F_y = \frac{d}{dt}(mv) \tag{2.4}$$

$$F_z = \frac{d}{dt}(mw) \tag{2.5}$$

Moment equation

$$L = \frac{d}{dt}(H_x) \tag{2.6}$$

$$M = \frac{d}{dt}(H_y) \tag{2.7}$$

$$N = \frac{d}{dt}(H_z) \tag{2.8}$$

Let δm be the element of mass of the bullet, v is the velocity relative to incritical absolute frame.

$$\delta F = \delta m \frac{dv}{dt}$$

(2.9)

total force exerting

$$\sum \delta f = F$$

(2.10)

Velocity of elementary mass

$$V = V_c + \frac{dr}{dt}$$

(2.11)

where, V_c = velocity of centre of mass.

$\frac{dr}{dt}$ = velocity of element relative to the centre of mass.

$$\sum \delta f = F = \frac{d}{dt} \sum (V_c + \frac{dr}{dt}) \delta m$$

(2.12)

Since mass of the bullet is constant

$$F = m \frac{dV_c}{dt} + \frac{d}{dt} \sum \frac{dr}{dt} \delta m$$

(2.13)

$$F = m\frac{dV_c}{dt} + \frac{d^2}{dt}\sum r \,\delta m$$

(2.14)

r is measured from centre of mass so the summation $\sum r \,\delta m$ is equal to zero.

The force equation thus becomes:

$$F = m\frac{dVc}{dt}$$

(2.15)

Now the momentum equation will be

$$\delta M = \frac{d}{dt}\delta H = \frac{d}{dt}(r*v)\,\delta m$$

(2.16)

and

$$V = V_c + \frac{dr}{dt} = V_c + w + r$$

(2.17)

where ω = angular velocity of bullet

r = position of element for centre of mass.

$$H = \sum \delta H = \sum (r*Vc)\,\delta m + \sum [r*(w*r)]\,\delta m$$

(2.18)

$$H = \sum \delta H = \sum (r*\delta m)\,Vc + \sum [r+(w*r)]\,\delta m$$

(2.19)

Angular velocity and position vector

$$w = pi + qi + rk \tag{2.20}$$

$$r = xi + yi + zk \tag{2.21}$$

$$H = (pi + qi + rk)\sum(x^2 + y^2 + z^2)\delta m - \sum(xi + yi + zk)(px + qy + rz)\delta m \tag{2.22}$$

Scalar components of H are

$$H_x = p\sum(y^2 + z^2)\delta m - q\sum(xy)\delta m - r\sum(xz)\delta m \tag{2.23}$$

$$H_y = -p\sum(xy)\delta m + q\sum(x^2 + z^2)\delta m - r\sum(yz)\delta m \tag{2.24}$$

$$H_z = -p\sum(xz)\delta m - q\sum(yz)\delta m + r\sum(x^2 + y^2)\delta m \tag{2.25}$$

Summation of this equation is product of interia of Bullet

$$I_x = \iiint (y^2 + z^2)\delta m \tag{2.26}$$

$$I_y = \iiint (x^2 + z^2)\delta m \tag{2.27}$$

$$I_z = \iiint (x^2 + y^2)\delta m \tag{2.28}$$

$$I_{xy} = \iiint (xy)\delta m \tag{2.29}$$

$$I_{xz} = \iiint (xz)\delta m \tag{2.30}$$

$$I_{yz} = \iiint (yz)\delta m \tag{2.31}$$

Larger the moments of interia, the greater will be the resistance to rotation.

$$H_x = pI_x - qI_{xy} - rI_{xz} \tag{2.32}$$

$$H_y = -pI_{xy} + qI_y - rI_{yz} \tag{2.33}$$

$$H_z = -pI_{xz} - qI_{yz} - rI_z \tag{2.34}$$

Now let have a arbitrary vector A referred to a rotating body frame with angular velocity ω.

$$\frac{dA}{dt}\bigg|_{Intialframe} = \frac{dA}{dt}\bigg|_{Bodyframe} + \omega * A \tag{2.35}$$

$$F = m\frac{dV_c}{dt}\bigg|_{Bodyframe} + m(\omega * V_c) \tag{2.36}$$

$$M = \frac{dH}{dt}\bigg|_{Bodyframe} + (\omega * H) \tag{2.37}$$

The scalar equation

$F_x = m(\dot{u} + qw - rv)$ (2.38)

$F_y = m(\dot{v} + ru - pw)$ (2.39)

$F_z = m(\dot{w} + pv - qu)$ (2.40)

$L = \dot{H}_x + qH_z - rH_y$ (2.41)

$M = \dot{H}_y + rH_x - pH_z$ (2.42)

$N = \dot{H}_z + pH_y - qH_x$ (2.43)

2.2 Orientation and position of Bullet

The orientation of a bullet can be described by three Consecutive rotations. The angular rotation are called Euler rotations.

Consider the bullet to be positioned so that the body axis system is parallel to fixed frame and the applying the following rotations:-

1. Rotate the x_f, y_f, z_f frame about Oz_f through the yaw angle to the frame to x_1, y_1, z_1.

2. Rotate the x_1, y_1, z_1 frame about Oy_1 through the pitch angle Θ bringing frame to x_2, y_2, z_2.

3. Rotate x_2, y_2, z_2 frame about Ox_2 roll angle Θ to x_3, y_3, z_3.

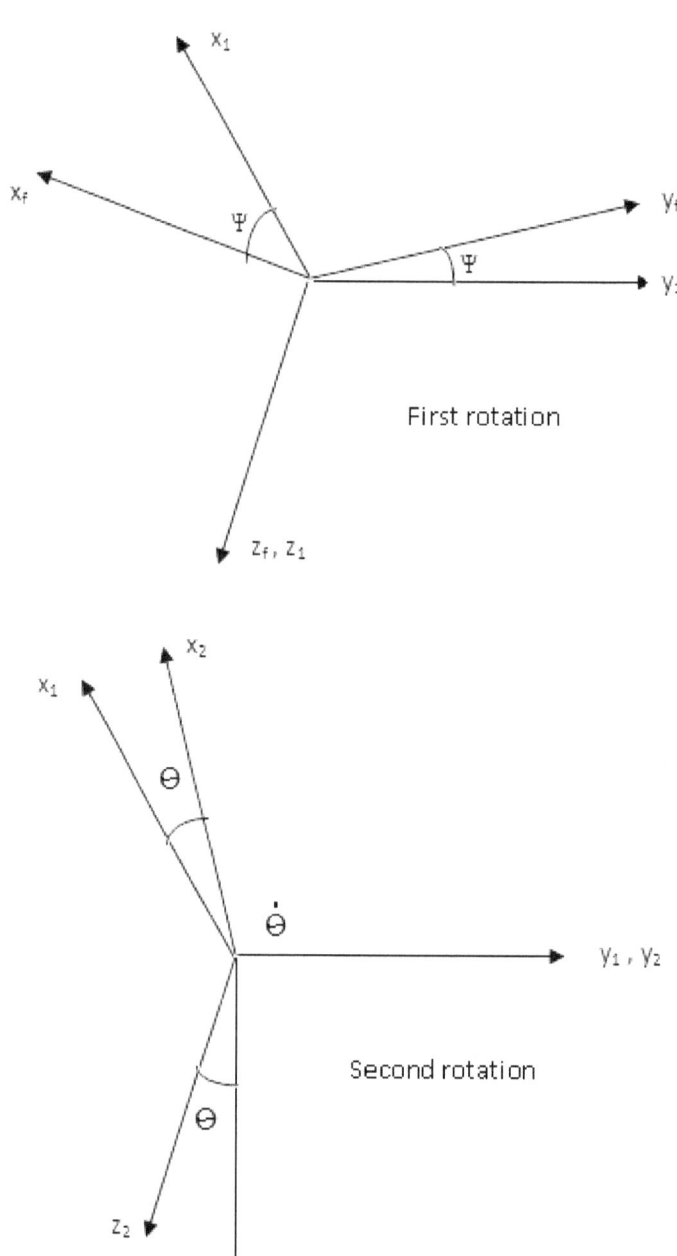

First rotation

Second rotation

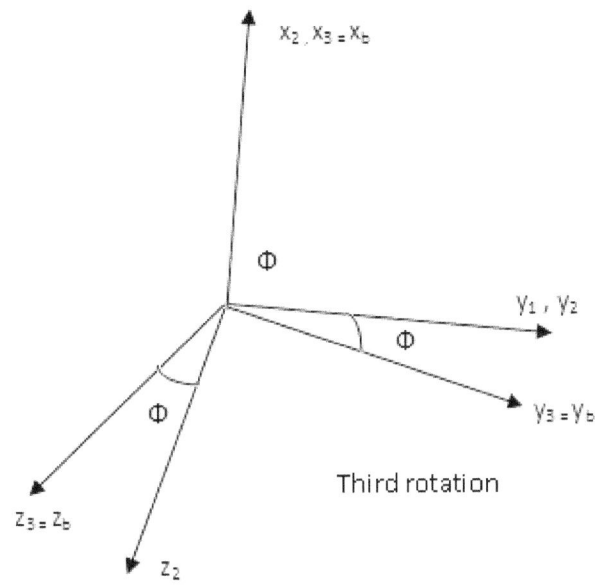

Figure no 2.2: Rotation/ Euler Angles

So from fig

$$\frac{dx}{dt} = u_1 \cos\Psi - v_1 \sin\Psi \tag{2.44}$$

$$\frac{dy}{dt} = u_1 \sin\Psi + v_1 \cos\Psi \tag{2.45}$$

$$\frac{dz}{dt} = w_1 \tag{2.46}$$

Let

$S_\Psi = \sin \Psi$ (2.47)

$C_\Psi = \cos \Psi$ (2.48)

$S_\Theta = \sin \Theta$

$C_\Theta = \cos \Theta$

(2.50)

$U_1 = U_2 C_\Theta + W_2 S_\Theta$

(2.51)

$V_1 = V_2$

(2.52)

$W_1 = -U_2 S_\Theta + W_2 C_\Theta$

(2.53)

$$\begin{bmatrix} \dfrac{dx}{dt} \\ \dfrac{dy}{dt} \\ \dfrac{dz}{dt} \end{bmatrix} = \begin{bmatrix} C_\Theta C_\Psi & S_\phi S_\Theta C_\Psi - C_\phi S_\Psi & C_\phi S_\Theta C_\Psi + S_\phi S_\Psi \\ C_\Theta S_\Psi & S_\phi S_\Theta S_\Psi - C_\phi C_\Psi & C_\phi S_\Theta S_\Psi - S_\phi C_\Psi \\ -S_\Theta & S_\phi C_\Theta & C_\phi C_\Theta \end{bmatrix} \begin{bmatrix} U \\ V \\ W \end{bmatrix}$$

(2.54)

For angular velocity

$$\begin{pmatrix} p \\ q \\ r \end{pmatrix} = \begin{pmatrix} 1 & 0 & -S_\Theta \\ 0 & C_\phi & C_\Theta S_\phi \\ 0 & -S_\phi & C_\Theta C_\phi \end{pmatrix} \begin{pmatrix} \dot{\phi} \\ \dot{\Theta} \\ \dot{\Psi} \end{pmatrix}$$

(2.55)

Force equation of Bullet

$$X - mg\, S_\Theta = m(\dot{u} + qw - rv)$$

$$Y + mg\, C_\Theta S_\phi = m(\dot{v} + ru - pw)$$

$$Z + mg\, C_\phi C_\Theta = m(\dot{w} + pv - qu)$$

(2.56)

Moment equation

$$L = I_x \dot{p} - I_{xz}\dot{r} + qr(I_z - I_y) - I_{xz}pq$$

$$M = I_y \dot{q} + rq(I_x - I_z) + I_{xz}(p^2 - r^2)$$

$$N = -I_{xz}\dot{p} + I_z \dot{r} + pq(I_y - I_x) + I_{xz}qr$$

(2.57)

Body angular velocities

$$p = \dot{\phi} - \dot{\Psi} S_\Theta \quad q = \dot{\Theta} C_\phi + \dot{\Psi} C_\Theta S_\phi$$

(2.58)

$$r = \dot{\Psi} C_\phi C_\phi - \dot{\Theta} S_\phi$$

Euler rates

$$\dot{\Theta} = q C_\phi - r S_\phi$$

$$\dot{\phi} = p + q S_\phi T_\Theta + r C_\phi T_\Theta$$

$$\tag{2.59}$$

$$\dot{\Psi} = (q S_\phi + r C_\phi) Sec_\phi$$

Velocity

$$\begin{matrix} \dfrac{dx}{dt} \\ \dfrac{dy}{dt} \\ \dfrac{dz}{dt} \end{matrix} = \begin{bmatrix} C_\Theta C_\Psi & S_\phi S_\Theta C_\Psi - C_\phi S_\Psi & C_\Psi S_\Theta C_\Psi + S_\phi S_\Psi \\ C_\Theta S_\Psi & S_\phi S_\Theta S_\Psi - C_\phi C_\Psi & C_\phi S_\Theta S_\Psi - S_\phi C_\Psi \\ -S_\Theta & S_\phi C_\Theta & C_\phi C_\Theta \end{bmatrix} \begin{matrix} U \\ V \\ W \end{matrix}$$

$$\tag{2.60}$$

CHAPTER 3

3.1 Overview of M16 Rifle

The bullet which we have taken in our project is of M16. The M16 is the United States military designation for the AR-15 rifle. M16 rifle fires the 5.56*45mm cartridge. It is a light weighted, 5.56 mm and it is made of steel, 7075 aluminium alloy, composite plastics and polymer materials. M16 rifle are further classified into M16A1, M16A2/A3, M16A4 etc.

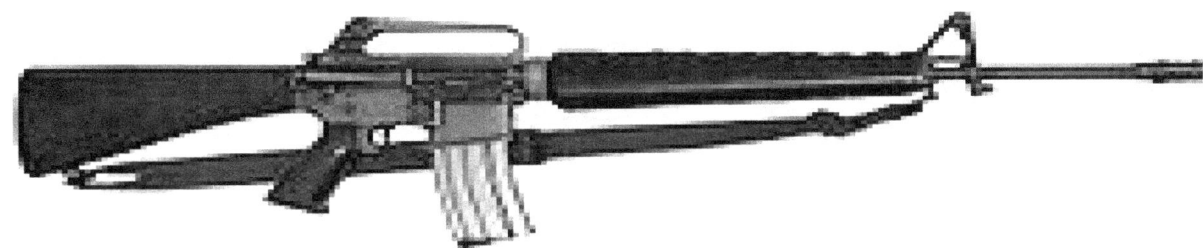

Fig No.3.1 M16 Rifle adapted from Url-3

The overview of different types of M16 rifles are:-

CHARACTERISTIC	M16A1	M16A2/A3	M16A4	M4
WEIGHT (pounds):				
Without magazine and sling	6.35	7.78	9.08	6.49
With sling and loaded:				
20-round magazine	6.75	8.48	9.78	7.19
30-round magazine	7.06	8.79	10.09	7.50
Bayonet knife, M9	1.50	1.50	1.50	1.50
Scabbard	0.30	0.30	0.30	0.30
Sling, M1	0.40	0.40	0.40	0.40
LENGTH (inches):				
Rifle w/bayonet knife	44.25	44.88	44.88	N/A
Overall rifle length	30.00	39.63	39.63	N/A
Buttstock closed	N/A	N/A	N/A	29.75
Buttstock open	N/A	N/A	N/A	33.0
OPERATIONAL CHARACTERISTICS:				
Barrel rifling-right hand 1 twist (inches)	12	7	7	7
Muzzle velocity (feet per second)	3,250	3,100	3,100	2,970
Cyclic rate of fire (rounds per minute)	700-800	700-900	800	700-900
MAXIMUM EFFECTIVE RATE OF FIRE:				
Semiautomatic (rounds per minute)	45-65	45	45	45
Burst (3-round bursts) (rounds per minute)	N/A	90	90	90
Automatic (rounds per minute)	150-200	150-200 A3	N/A	N/A
Sustained (rounds per minute)	12-15	12-15	12-15	12-15
RANGE (meters):				
Maximum range	2,653	3,600	3,600	3,600
Maximum effective range				
Point target	460	550	550	500
Area target	N/A	800	600	600

3.2 Dimension of the M16 Cartridge

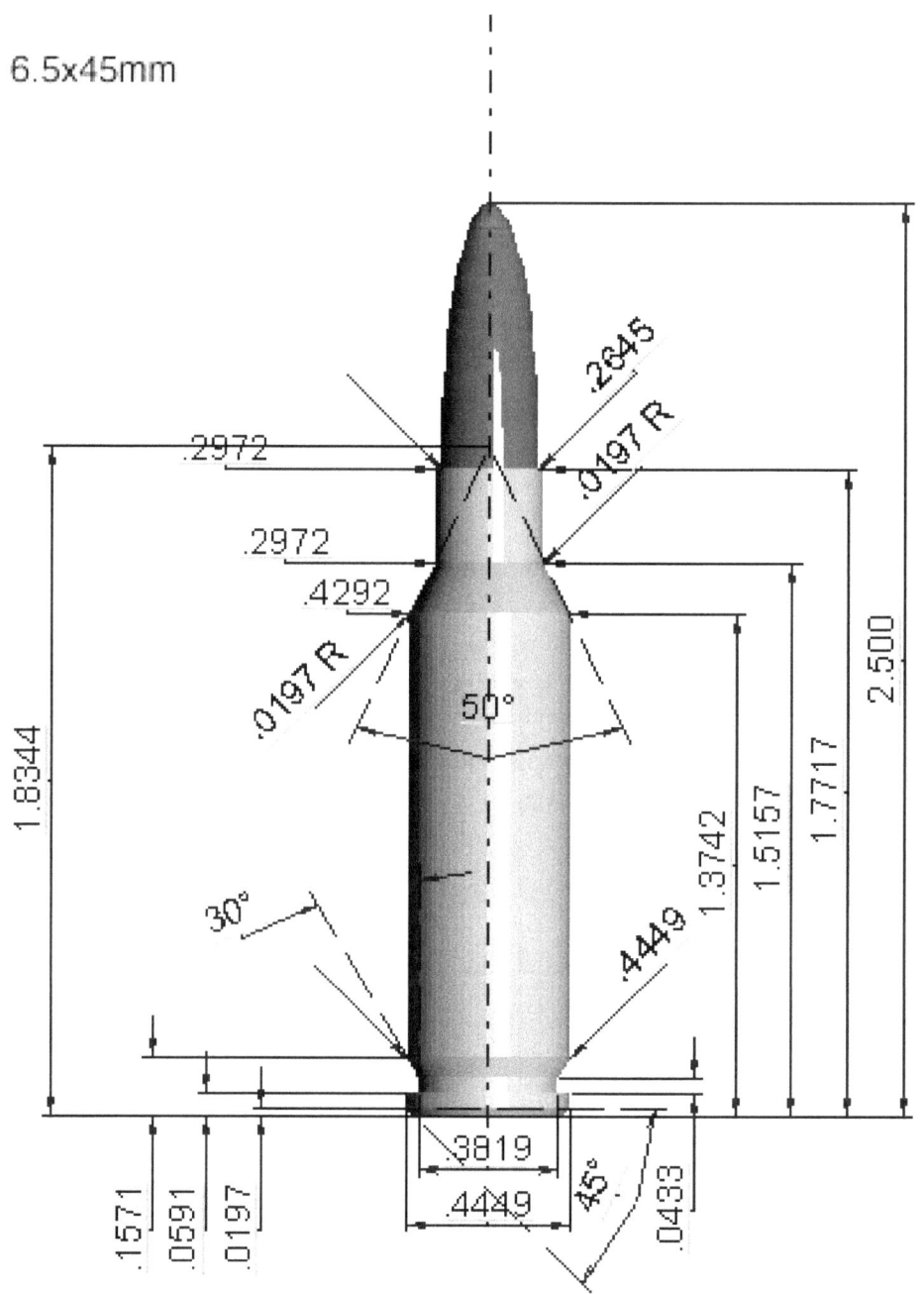

Fig.No. 3.2 adapted from Url-3

From the above figure we can see that the cartridge is divided into two different part the first one that is of red and brown colour is called bullet and the second part which is of wooden colour is called shell. The diameter of the bullet is 5.56 mm, and its length is 12.7 mm.

3.3 Types of M16 bullet

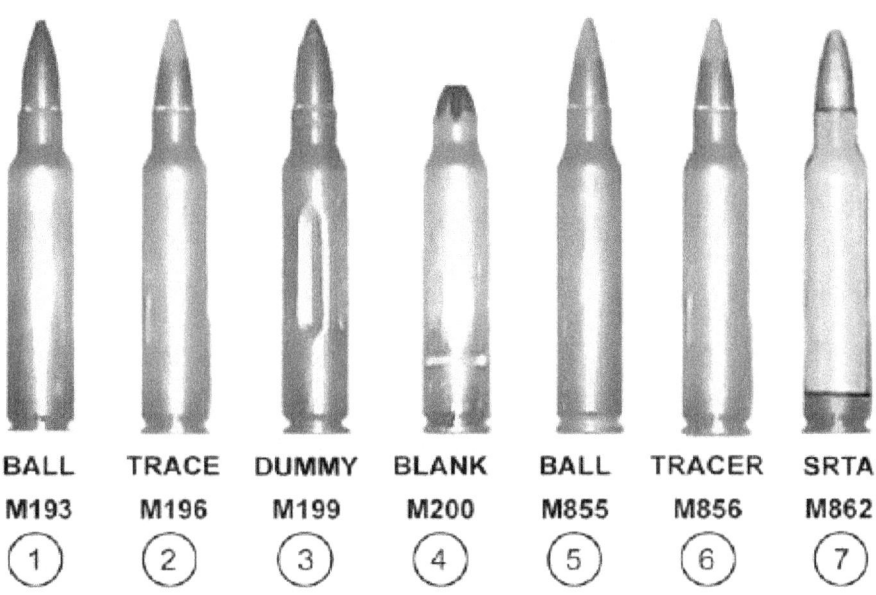

Fig.No 3.3 adapted from Url-3

1. Cartridge, Ball, M193:- It is a center-fire cartridge of 55 grain made of lead alloy.

2. Cartridge, Tracer, M196:- This cartridge has red or orange coloured tip. In long term use of this cartridge should be avoided as it can damage the barrel.

3. Cartridge, Dummy, M199:- During dry firing and in training program this cartridge is used. The best way to identify it is that it have six grooves along the side of the case beginning about ½ inch from the tip.

4. Cartridge, Blank, M200:- This cartridge is not used in that case where projectile are more required.

5. Cartridge, Ball, M855:- It is made up of lead alloy and has steel penetrator and is of 62 grain. Its tip is painted with green colour.

6. Cartridge, Tracer, M856:- This cartridge is of 63.7 grain. Its tip is of red colour.

7. Cartridge, Short-Range Training Ammunitin (SRTA), M862:- Main purpose of this bullet is that it is exclusively designed for training purpose.

CHAPTER 4

Aerodynamics behind the Projectile

4.1 Boundary layer

All kind of projectiles when projected in the fluid, will be encountered with a film generated between the projectile and free flowing fluid called as the boundary layer. At that layer, fluid flow is retarded due to the friction/viscosity between the solid surface and fluid. An invisible layer of gas or liquid is formed over the surface of the projectile depending on the velocity. This is because all kinds of mass transfer, heat transfer, momentum transfer, friction effects and all viscosity effects are felt by the bullet. Therefore, instead of solving the whole Navier stokes equation for the system, only its necessary to solve for the boundary layer where all kinds of transfers are felt.

There are two type of boundary layer:

1 Laminar boundary Layer

2 Turbulent boundary Layer

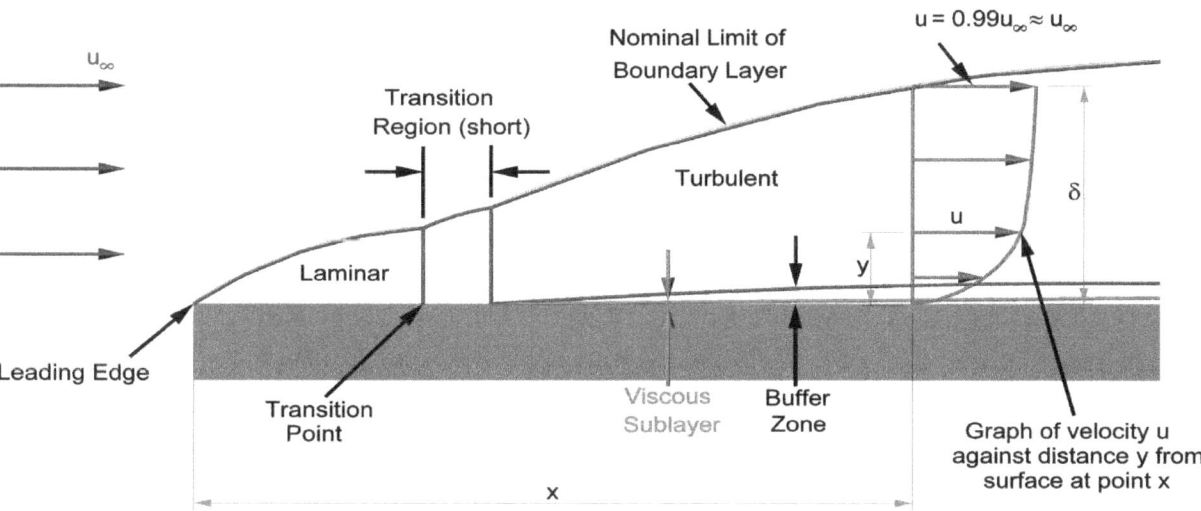

Figure 4.1 boundary layer adapted from Url-4

Laminar boundary layer is the boundary layer formed when flow is laminar on the contrary the turbulent boundary layer formed due to turbulent flow. In laminar flow, the streamline never intersects each other, but in turbulent flow the streamlines always intersect which shows the randomness of the fluid particles. Generally, the predication of laminar or turbulent flow can be done by Reynolds no. High Reynolds No. may show turbulence but the vice versa may not be always true.

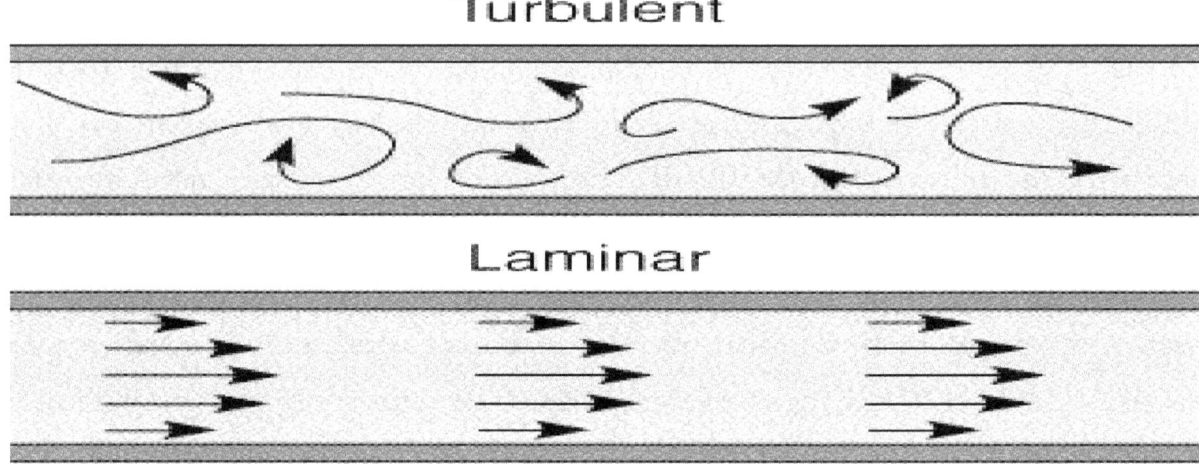

Figure no.4.2 Laminar and turbulent flow adapted from Url-4

The boundary layer formed increases drag and offers hindrance to the moving projectile.

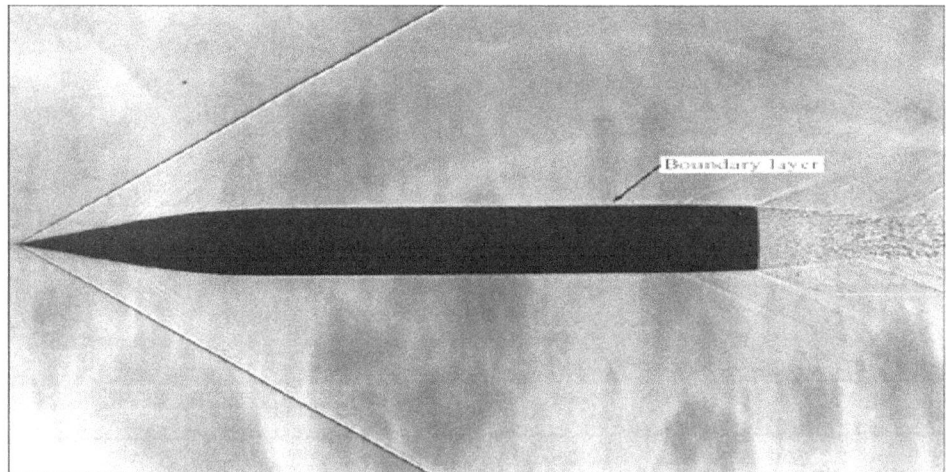

Figure No: 4.3 boundary layer over an ogive body

(Ref: Fundamental Aerodynamics J.D Anderson)

4.2 Separation of boundary layer

The boundary layer may get separated and form a new effective shape which is very much different than the original one. The separation of boundary layer occurs due to low energy relative to the free stream and easily lift off by change in pressure. It is bound to occur at large pressure gradients (increase of the pressure flow in the direction of the flow). Increase in fluid is very similar to the increase in the potential energy, thus this reduces the kinetic energy which results to declaration of projectile.

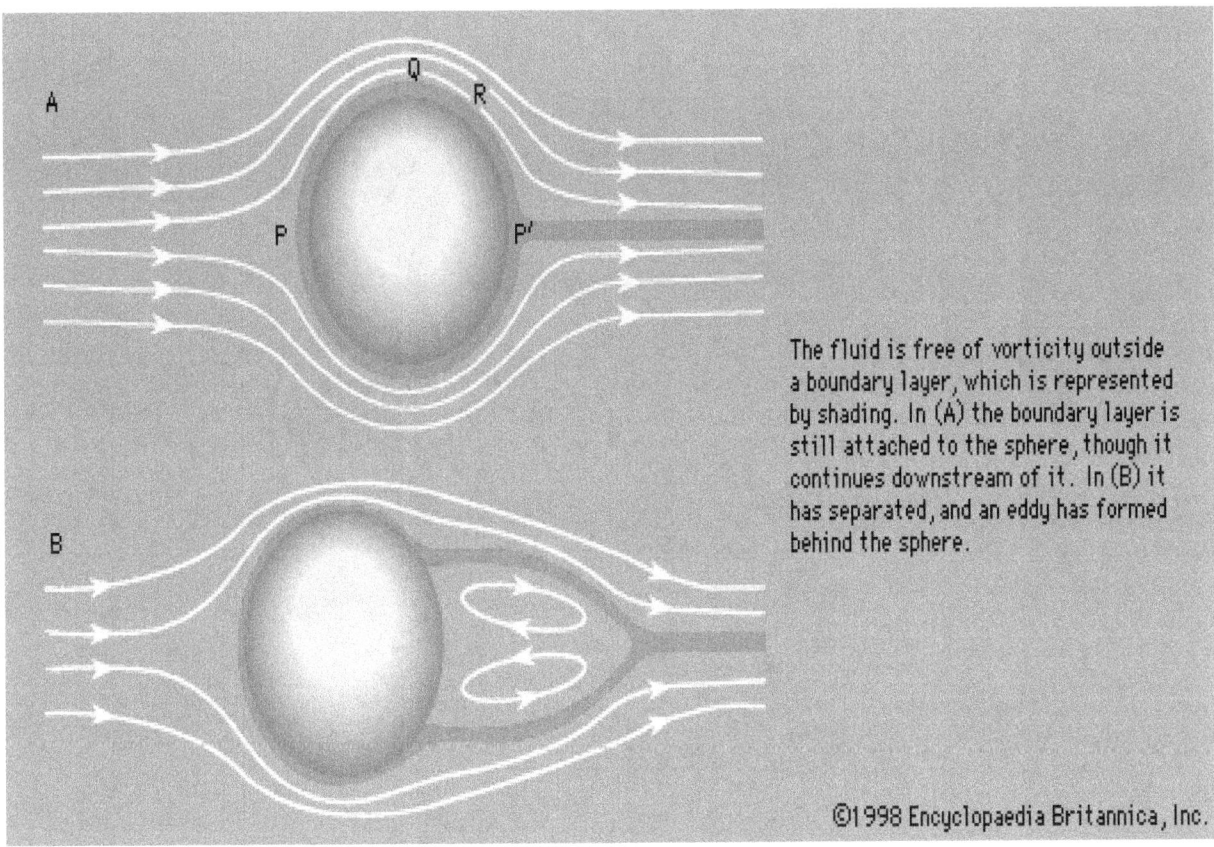

Figure.No. 4.4 boundary layer separation adapted from Url-6

4.3 Formation of shock waves

Since all type of flow are compressible in nature (density is not treated as constant), and such flows are high energy flow where major change in temperature and energy occurs. Whenever the speed of the projectile is faster than the speed of sound, than shock wave as projectile is in the compressible region.

A shockwave is a very thin region where the properties of flow change drastically. Its thickness is usually of order of 0.00001 cm where explosive compression occurs. When a projectile is encountered by a shock wave, the total energy remains conserved but the energy which can be extracted as work decreases, and therefore, entropy increases. Due to this, an additional drag force originates with shocks.

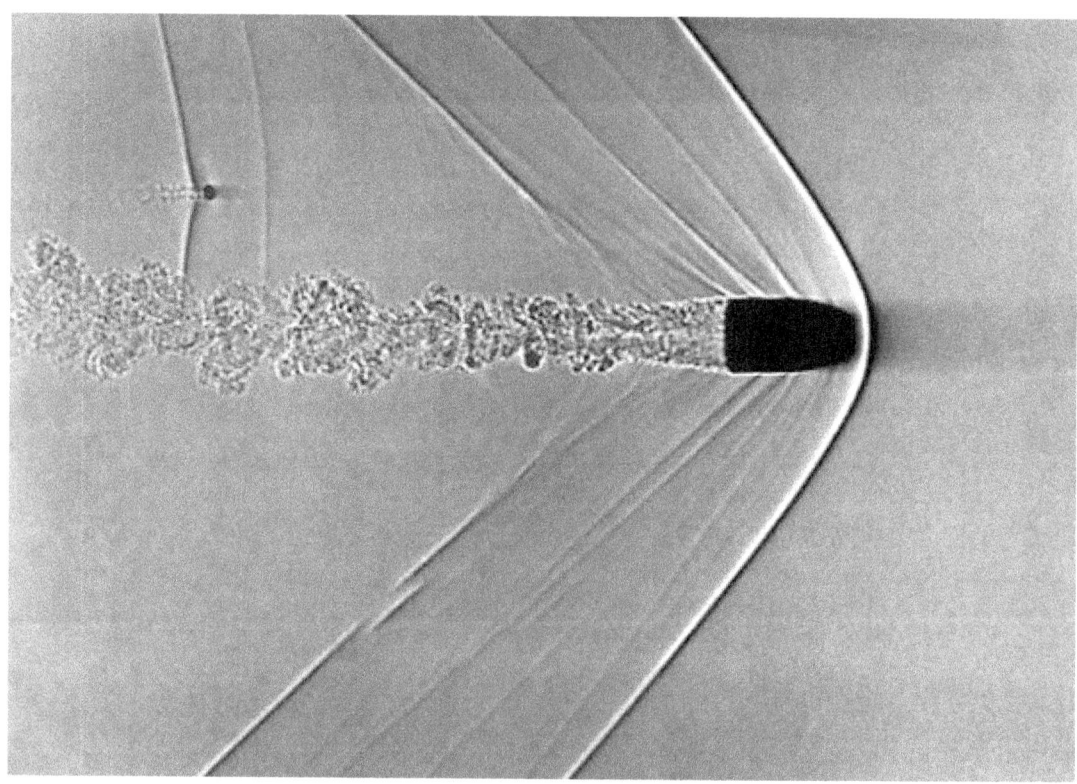

Figure. 4.5: Bow shock adapted from Url-5

The strength and type of shock depends on velocity of projectile. The shock wave generated at Mach No 1 (velocity of projectile equals to velocity of sound). At transonic range the projectile encountered by bow shock but at supersonic speed it encountered by oblique shock.

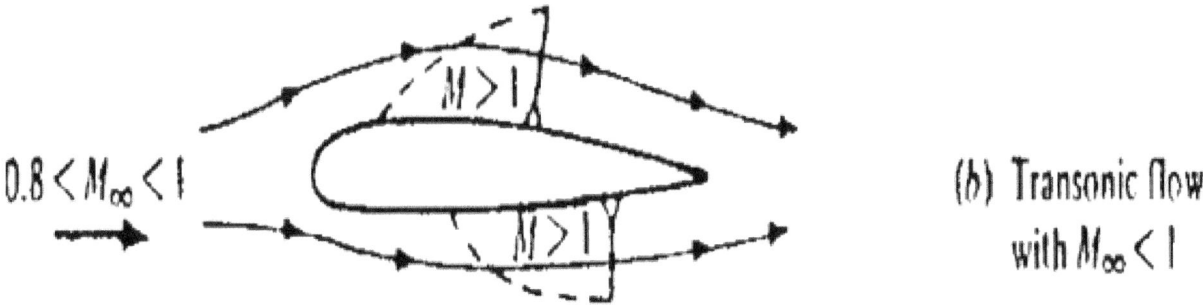

Figure.No.4.6 Transonic flow/lambda shock adapted from Url-6

At transonic flow, which is between 0.8 to 1, lambda shocks formed where flow is greater than Mach 1

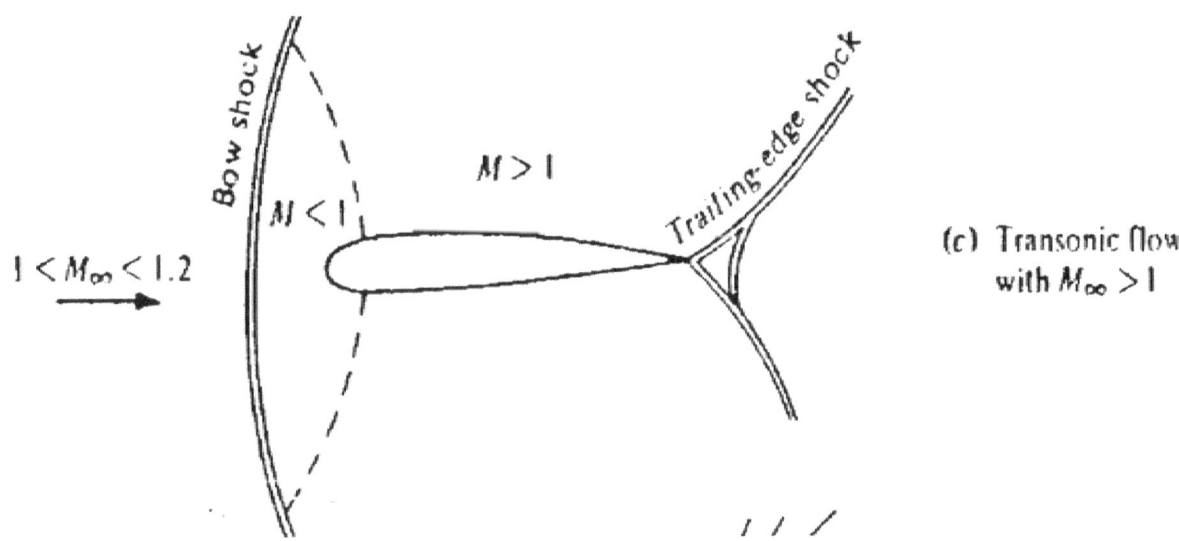

Figure.No. 4.7 Transonic flow /Bow shock adapted from Url-6

At Mach greater than 1 and when less than 1.2 a bow shock formed which detached from body.

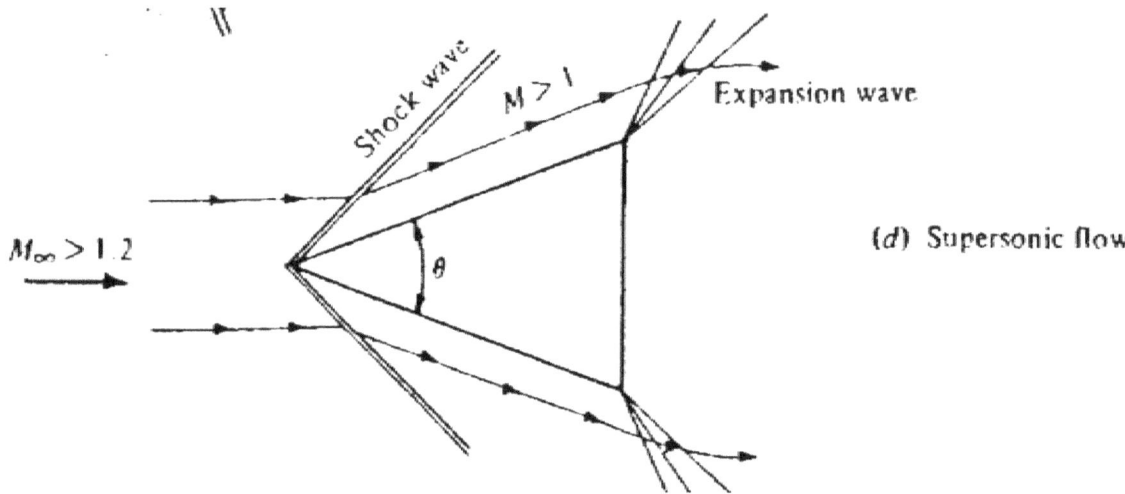

Figure.No.4.8 Supersonic flow/ oblique shock adapted from Url-6

When flow is at supersonic speed greater than Mach 1.2 oblique shock formed.

4.4 Change in properties across the Shock

The pressure, density, temperature, and entropy always increase across the shock, while Mach number and velocity decreases.

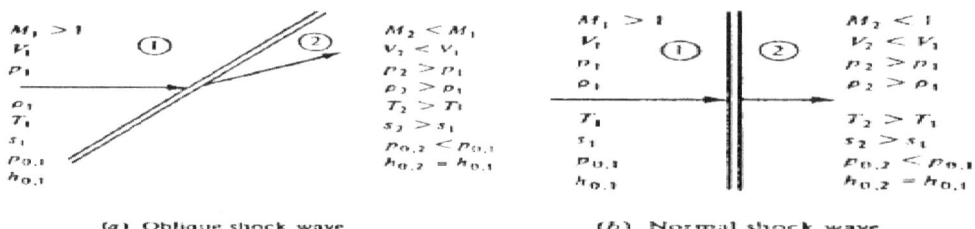

Figure.No.4.9 oblique Shock wave adapted from Url-6

Figure.No.4.10 Normal shock wave adapted from Url-6

CHAPTER 5

5.1 The main objective was to analyse the effect of flow over the bullet and show its advantage of its shape in terms of range and penetration. Therefore, in the chapter the 3 different shapes were taken and it was attempted to analyse the flow over the geometries. The geometries are:

1. Point Head Bullet
2. Pure blunt head bullet
3. Semi blunt head(original shape)

5.1.1 Point Head Bullet

The alteration has been done in the head of bullet, which is changed to point head from semi blunt shape. The design has been made in CATIA with no change in bore and length. The length of bullet is 12.7 mm with 5.56mm diameter.

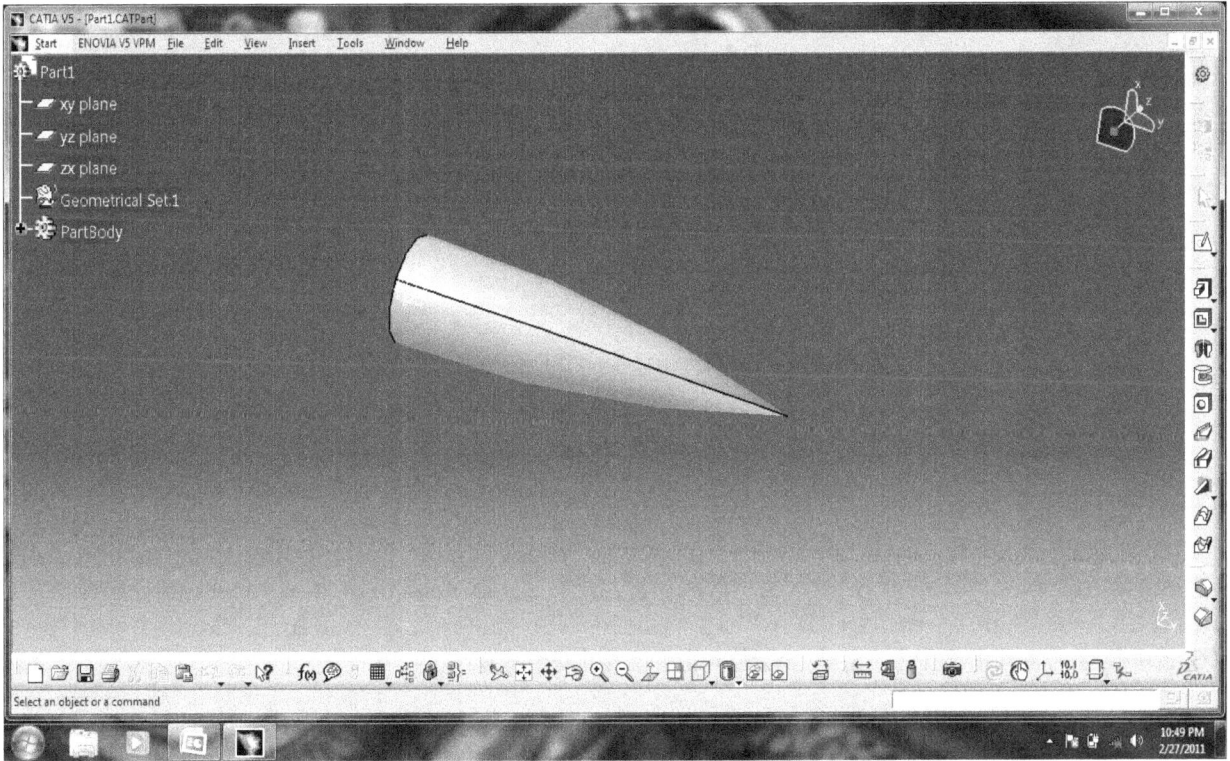

Figure.No 5.1 CATIA model of bullet

The initial measurement of the point head projectile is also measured in CATIA, such as centre of gravity, area, mass, moment of interia etc, which is displayed below.

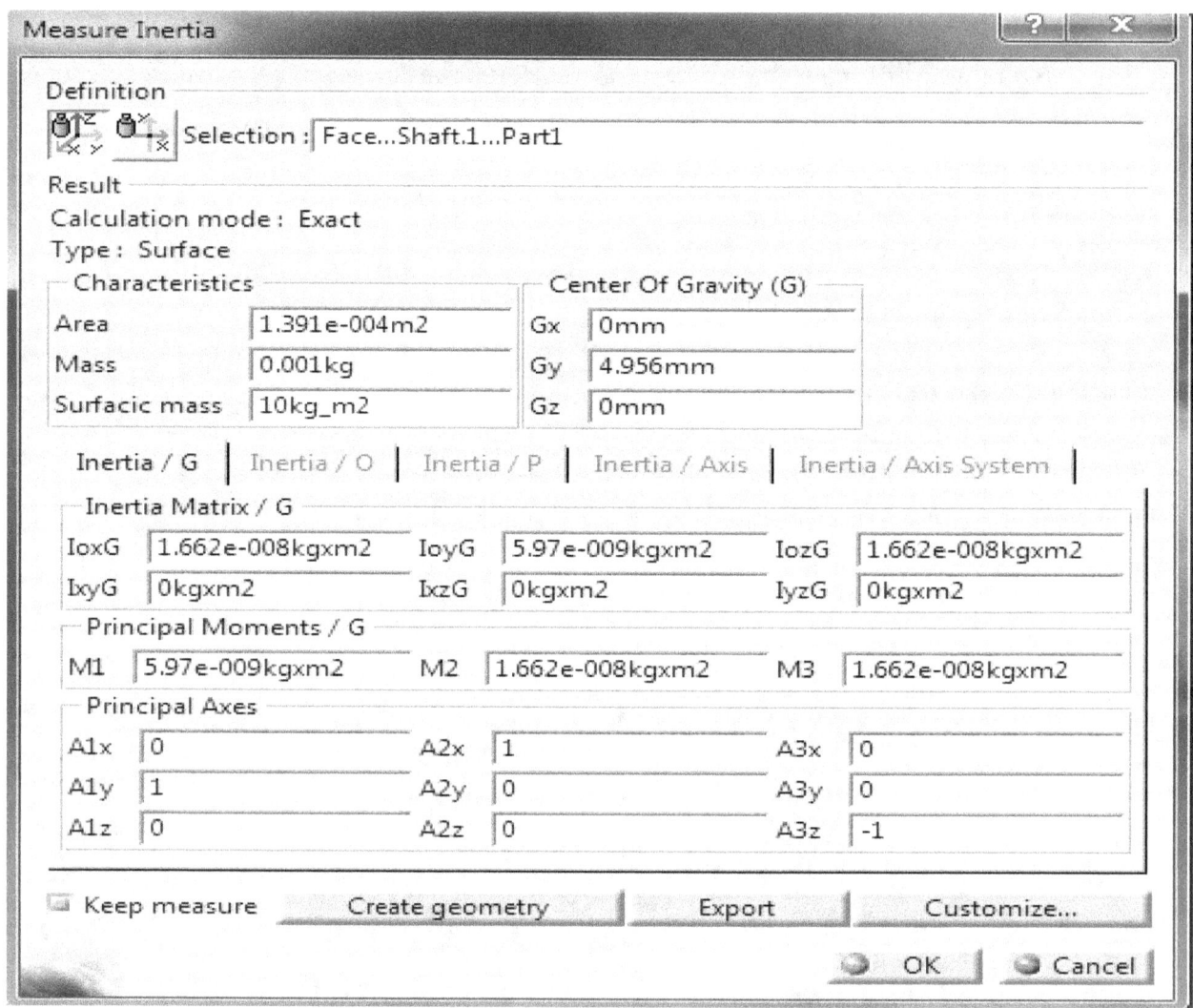

Figure.No.5.2 measure inertia of bullet in CATIA

The meshing has been done with the help of GAMBIT and for the analysis fluent has been used.

It has been tried to get selective meshing, which is fine on the bullet and near the bullet because the major change occurs at the bullet surface and near to it. For meshing GAMBIT is used and for analysis part FLUENT 6.3 is used. The material of bullet is taken as steel form fluent database which very close to the material of actual bullet.

The figure below is displaying the grid of point head projectile (M16 Bullet) and nearby surrounding.

Figure.No.5.3 Grid over a point head bullet

The figure below is showing dynamic pressure at supersonic speeds. Due to point head shape stagnation point is missing, therefore pressure is not distributed at the effective area. Shock waves are attached to the body.

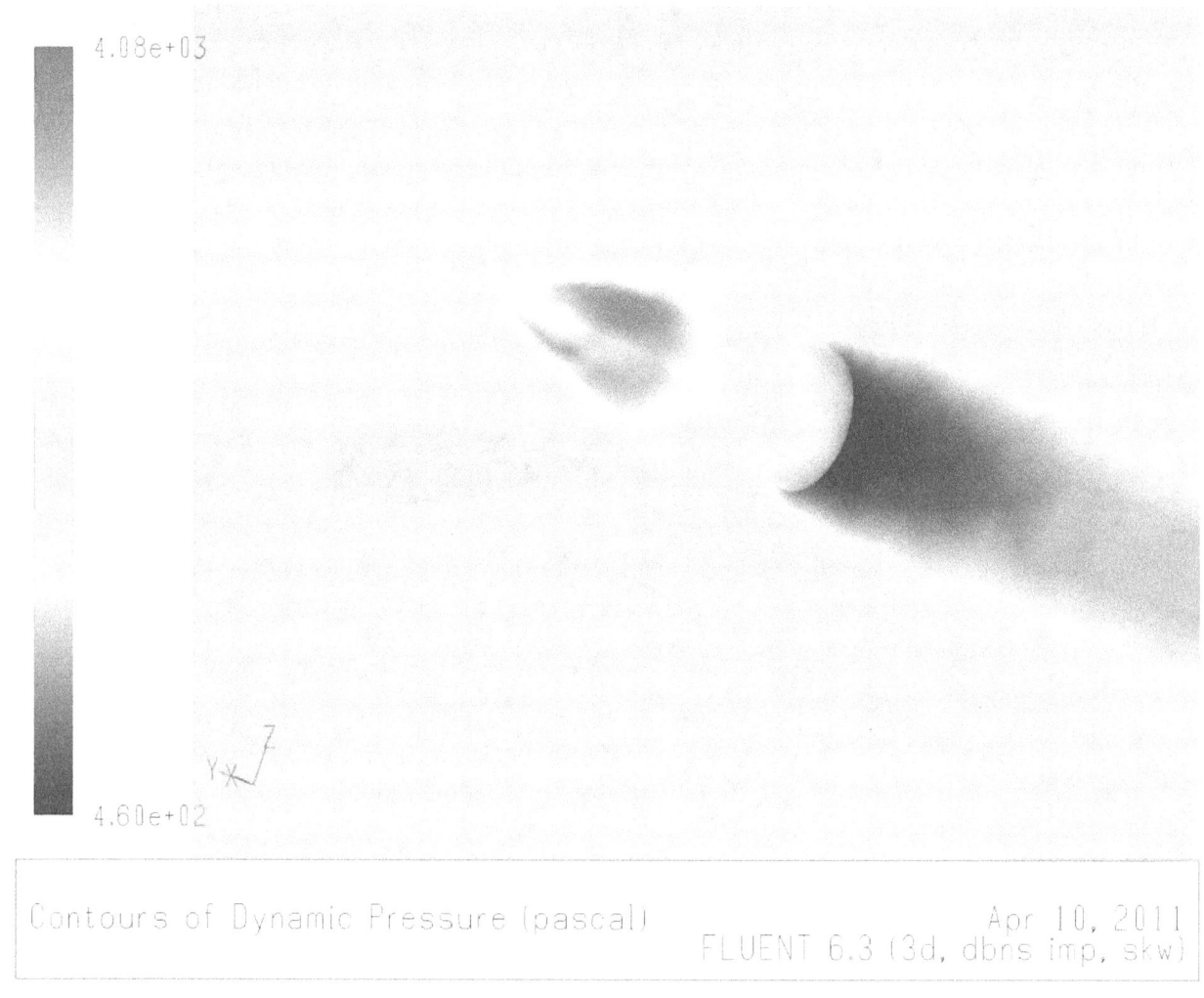

Figure.No.5.4 Contours of Dynamic pressure of point head bullet

Similarly, the figure below is showing the static pressure. The maximum pressure is exerting on the body rather than on point.

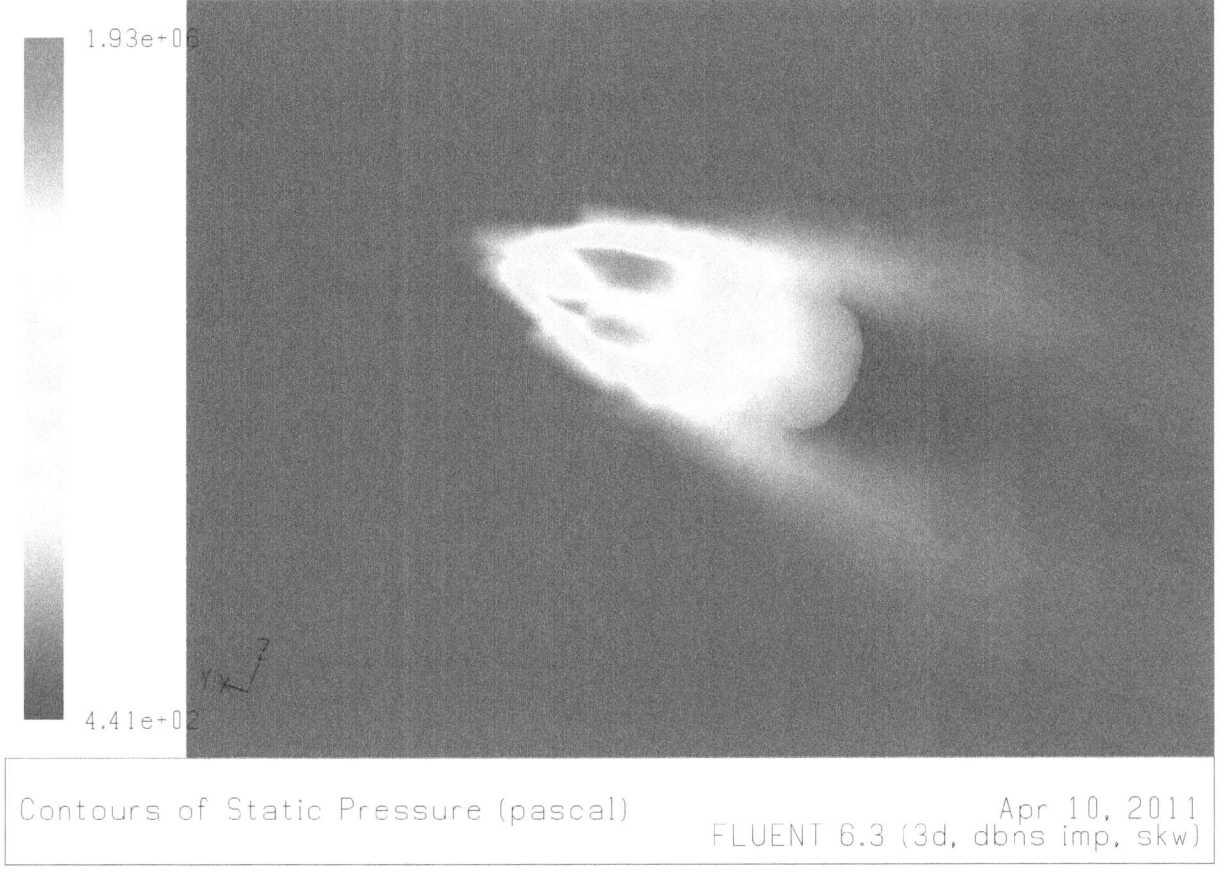

Figure.No.5.5 Contours of static pressure of point head bullet

At supersonic speeds for a pointed body, the boundary layer is attached to surface and increases the displacement thickness. As shown in figure below:

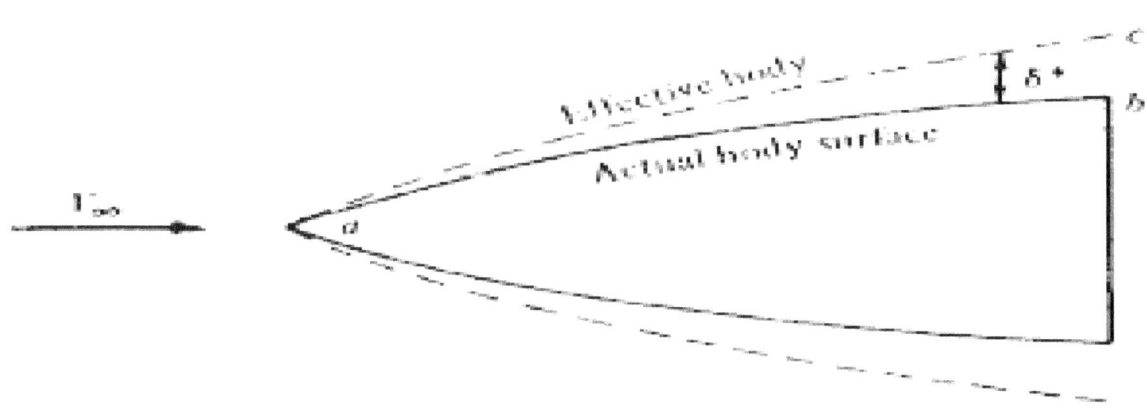

Figure.No, 5.6 Displacement thickness adapted from Url-6

Due to increase in the body surface area, the skin friction drag dominates and thus provides hindrance to the projectile.

5.1.2 Analysis for a pure cylindrical body

The design of the actual bullet is changed from head and converted to a pure blunt geometry. There is no change made in the diameter and length of the bullet. The material is also not changed. The figure below displays the meshed bullet and its surroundings.

Figure.No.5.7 Grid of a pure blunt head bullet

Below the contour, the display shows the static pressure, which is maximum at stagnation point thus temperature is also at its apex.

Due to the dome shape, the pressure is distributed over the surface of projectile. Thus, there is gradual decrement in the pressure.

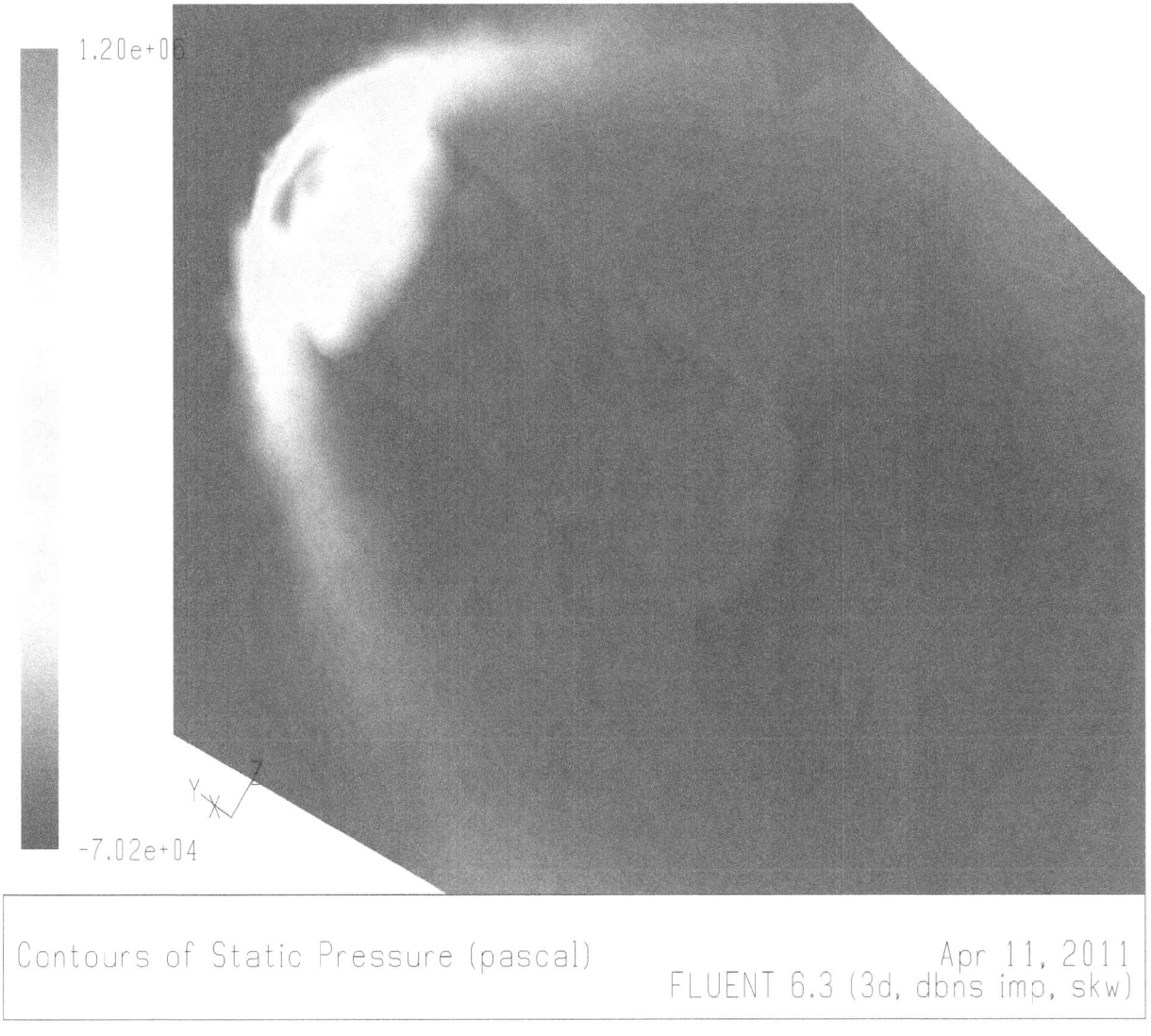

Figure.No.5.8 Contours of static pressure of pure blunt head bullet

The next contour is of dynamic pressure. At stagnation point, it is showing the lowest value, which is theoretically correct. The red region around the bullet is shock formation where sudden increment of pressure occurs. The dynamic pressure should be low at that point where static pressure is high.

Bow shock is stronger form of shock rather than oblique shock. The pressure increases very dramatically. The temperature and density also encounters with the same.

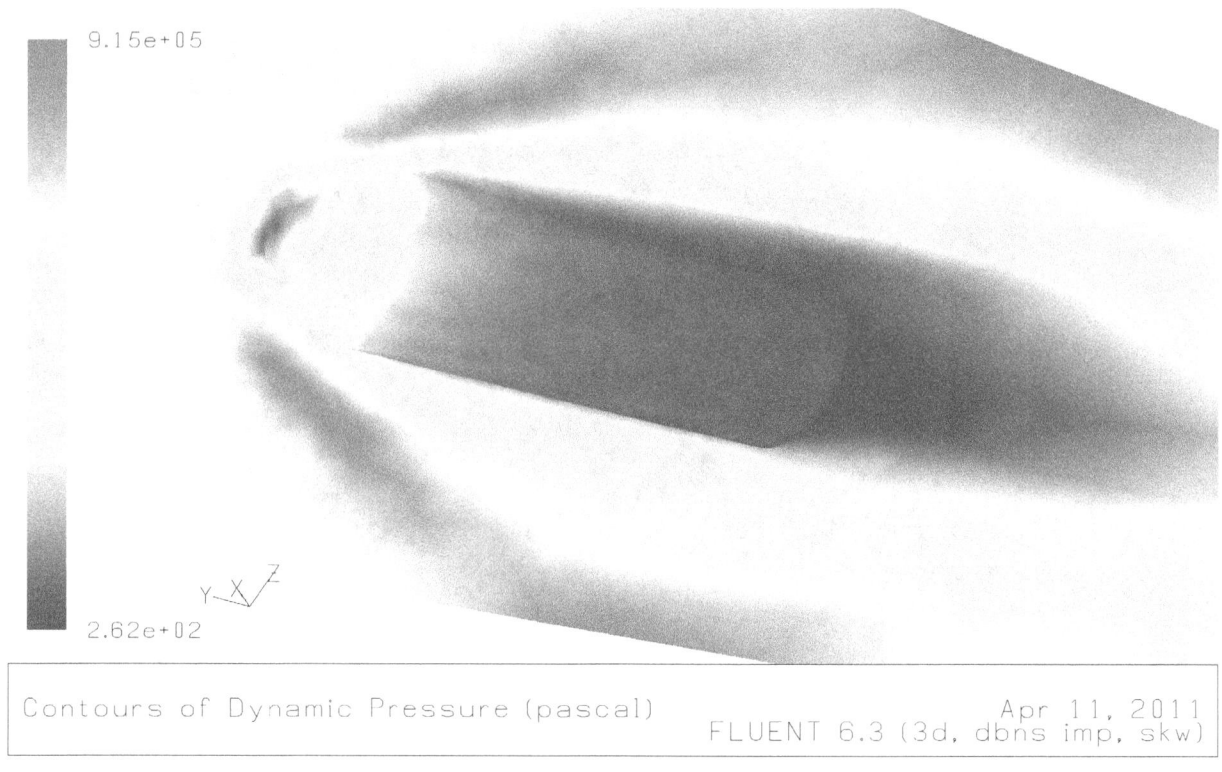

Figure.No.5.9 Contours of Dynamic pressure of pure blunt head bullet

The next contour shows the path lines:

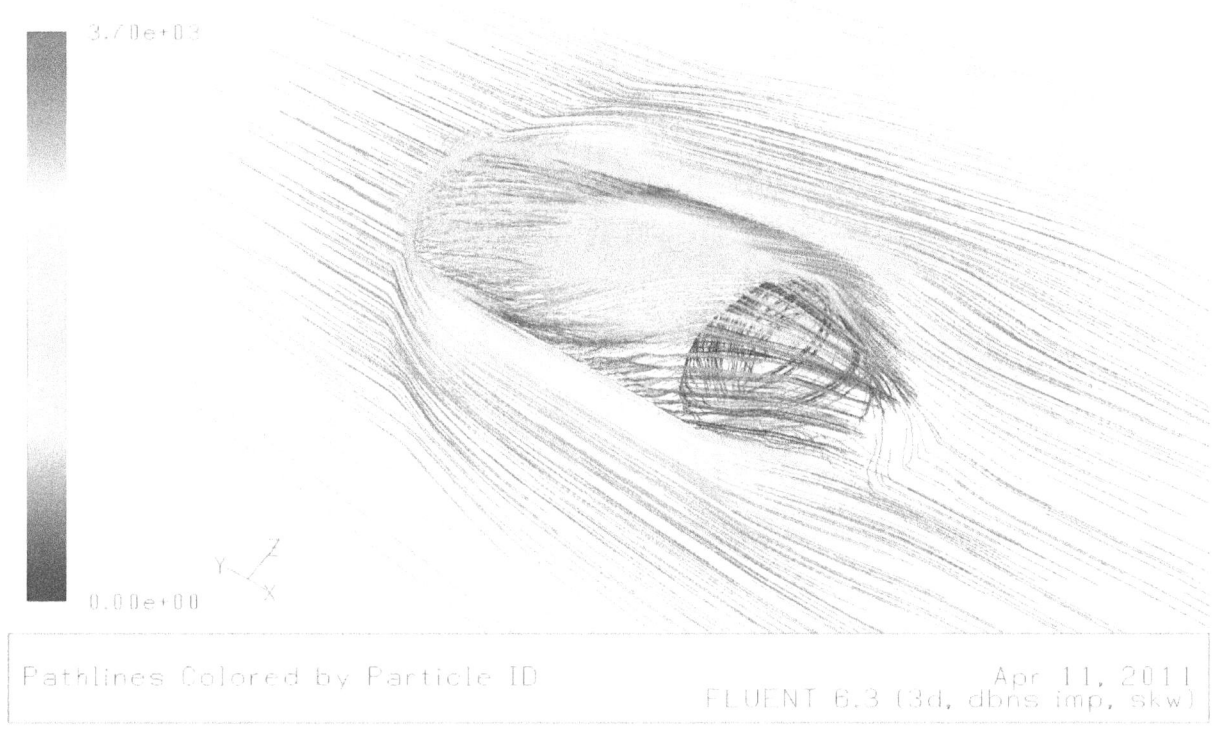

Figure.No.5.10 Path line of pure blunt head bullet

Since both the shape discussed above had particular disadvantages

This is the original shape of M 16 bullet. The length is 12.7mm and diameter 5.56mm.

CHAPTER 6

6.1 Analysis of Semi blunt head bullet at different angle of attack and Mach no.

For the analysis of the actual bullet, different Mach no at different angles of attack is taken in supersonic range.

- 6.1.1 Mach No 3.5
    - a. 0 degree angle of attack
    - b. 2 degree angle of attack
    - c. 2.5 degree angle of attack
- 6.1.2 Mach no 3
    - a. 0 degree angle of attack
    - b. 2 degree angle of attack

6.1.1. a. Mach no 3.5 0-degree angle of attack

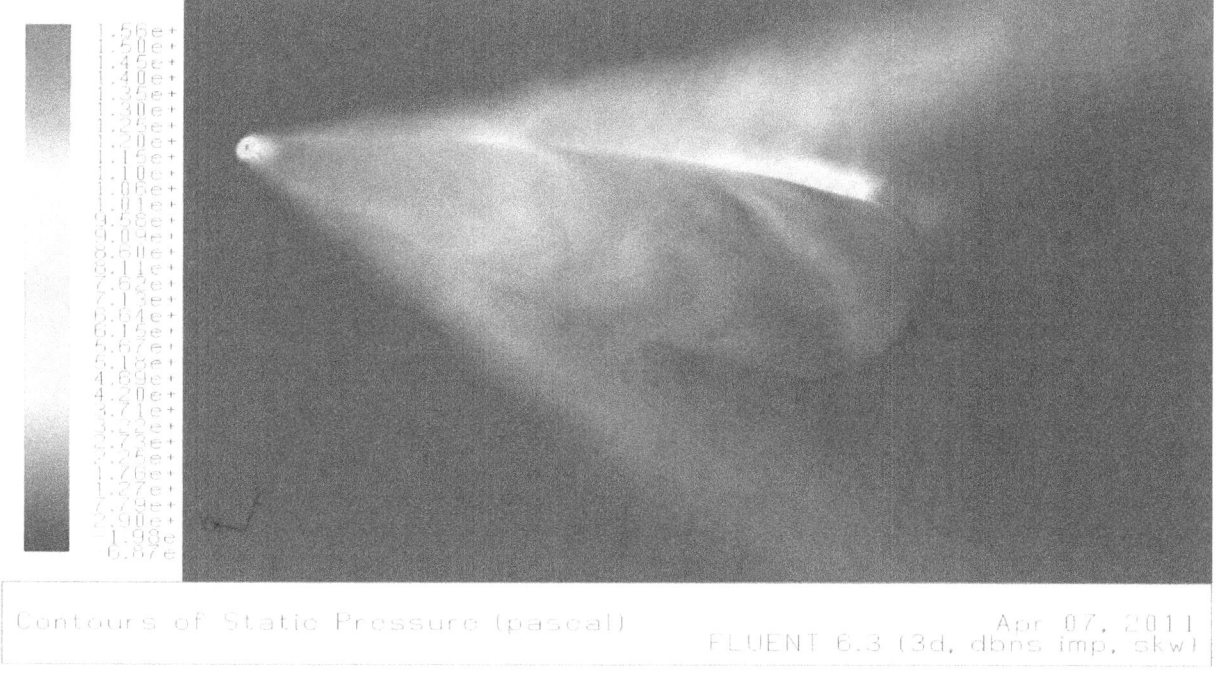

Figure.No.6.1 Contours of static pressure of semi blunt head bullet at 3.5 mach and zero angle of attack.

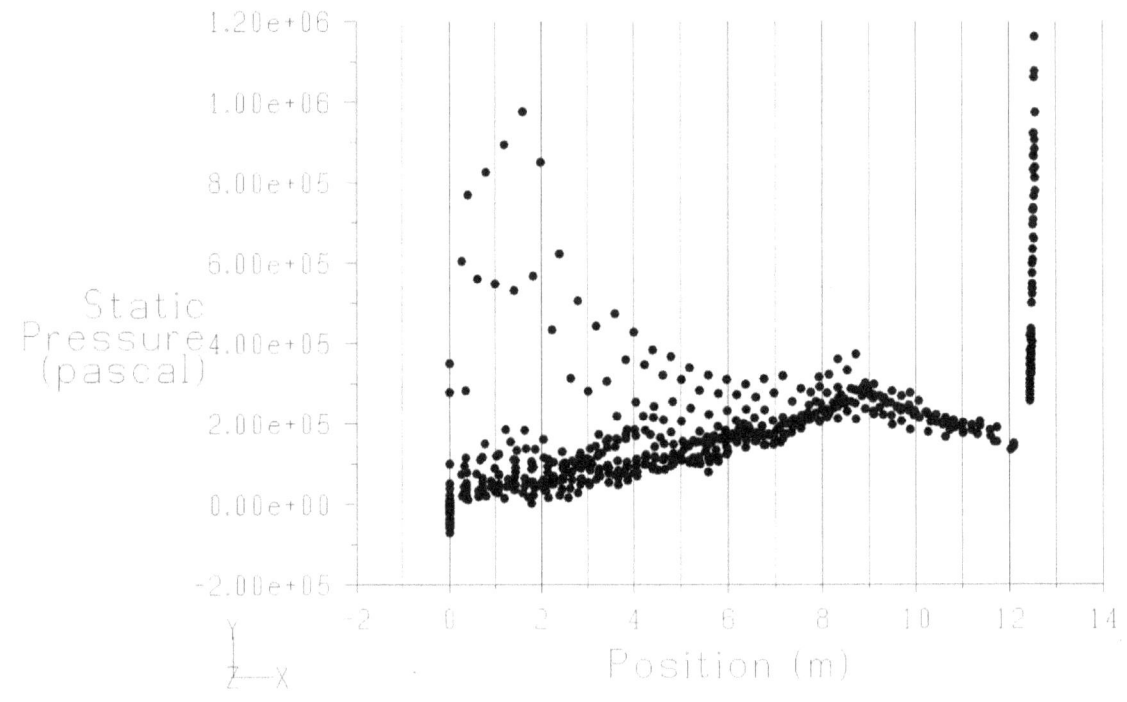

Graph.No.6.1 Static pressure graph

The above figures are the contour of static pressure and graph of static pressure of the projectile (M16 bullet) at 3.5 Mach at 0-degree angle of attack. The static pressure is maximum at stagnation point. It is 1.18e+06 pascal at stagnation point which is at 12.7 mm from the origin (length of bullet is 12.7mm). There is sudden drop in static pressure at 12mm and the value comes down to 2.00e+05 pascal. It is because at the stagnation point velocity is zero and therefore pressure is at its maximum value.

At 0 mm, the static pressure is very low, because flow gets turbulent due to the shape and it is observed that the velocity is very high there.

The figure below displays the dynamic pressure.

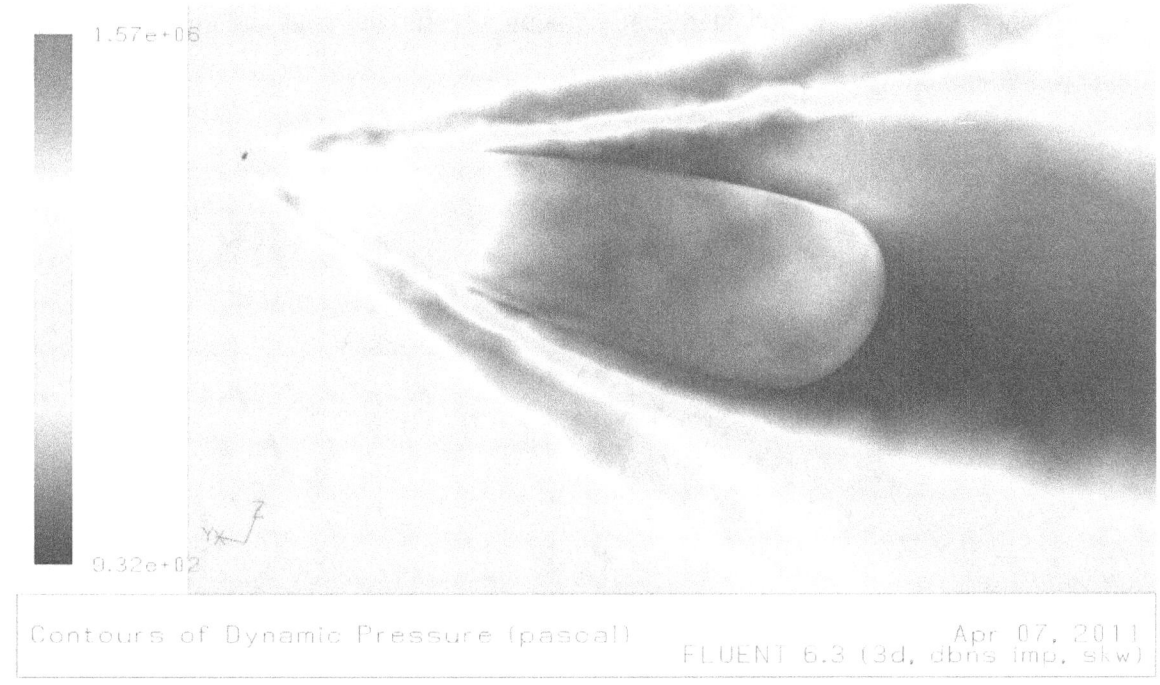

Figure.No.6.2 Contours of Dynamic pressure of semi blunt head bullet at 3.5 mach and zero angle of attack.

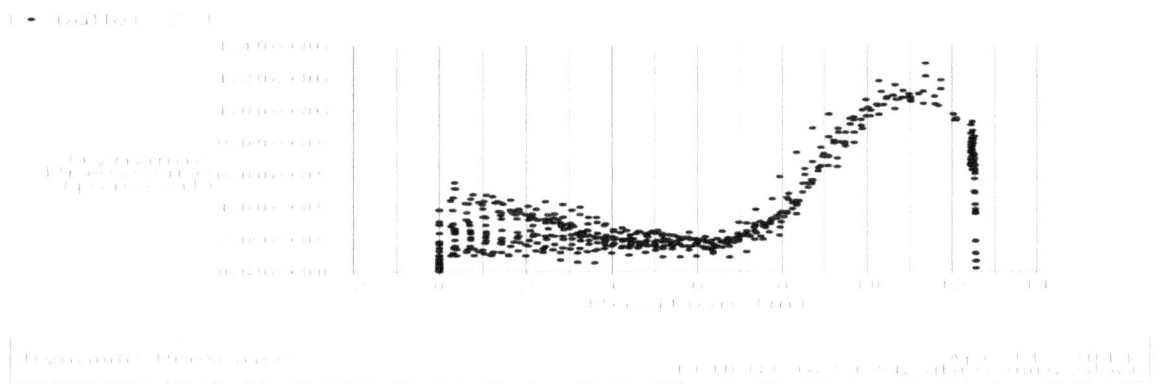

Graph.No.6.2 Dynamic pressure graph

The dynamic pressure is always low where static pressure is high. Maximum dynamic pressure goes till 1.3e+06 pascal. And it is almost zero at stagnation point.

6.1.1. b. 3.5 Mach 2-degree angle of attack

The figure beneath is displaying the static and dynamic pressure contour and graphs at 3.5 Mach at 2-degree angle of attack.

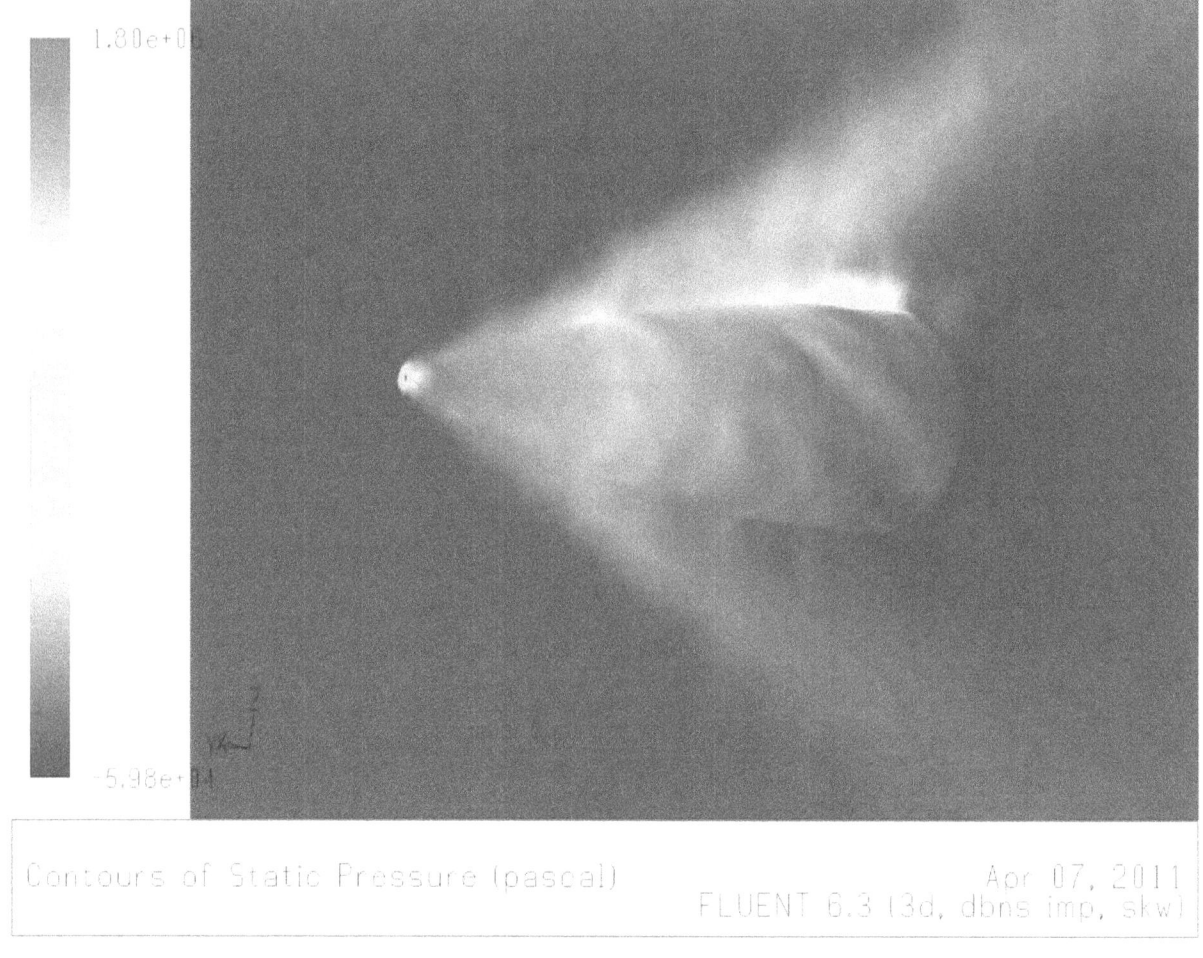

Figure.No.6.3 Contours of static pressure of semi blunt head bullet at 3.5 Mach and Two degree angle of attack

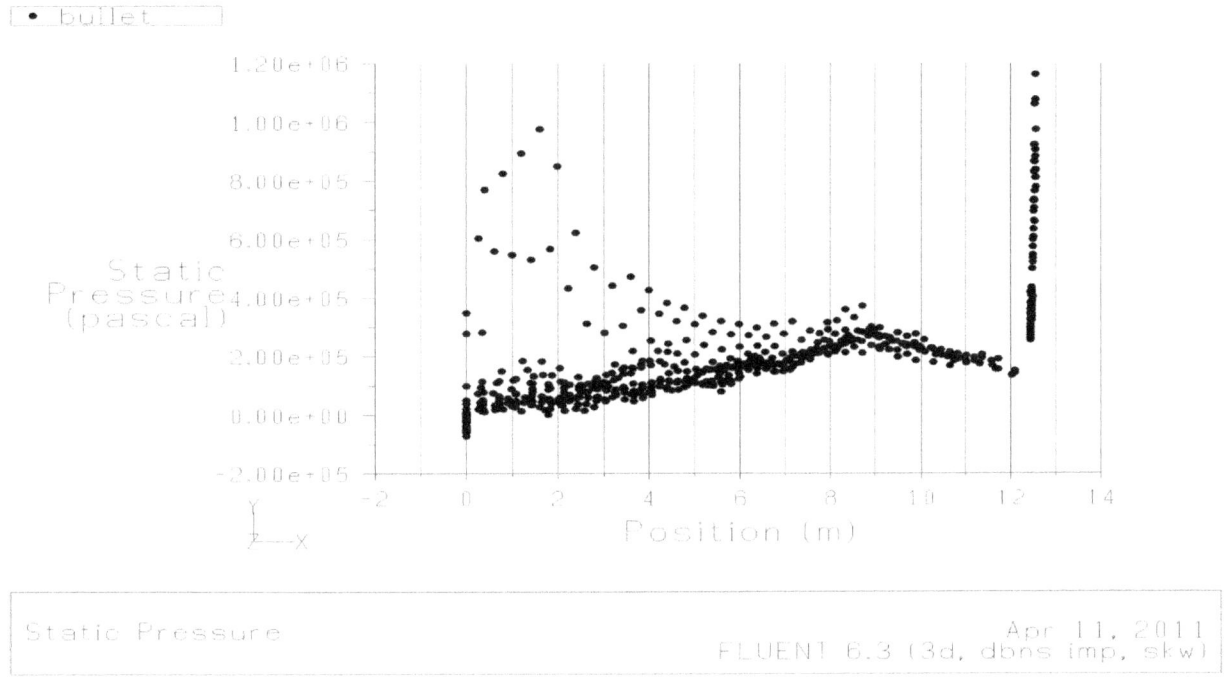

Graph.No.6.3 Static pressure graph

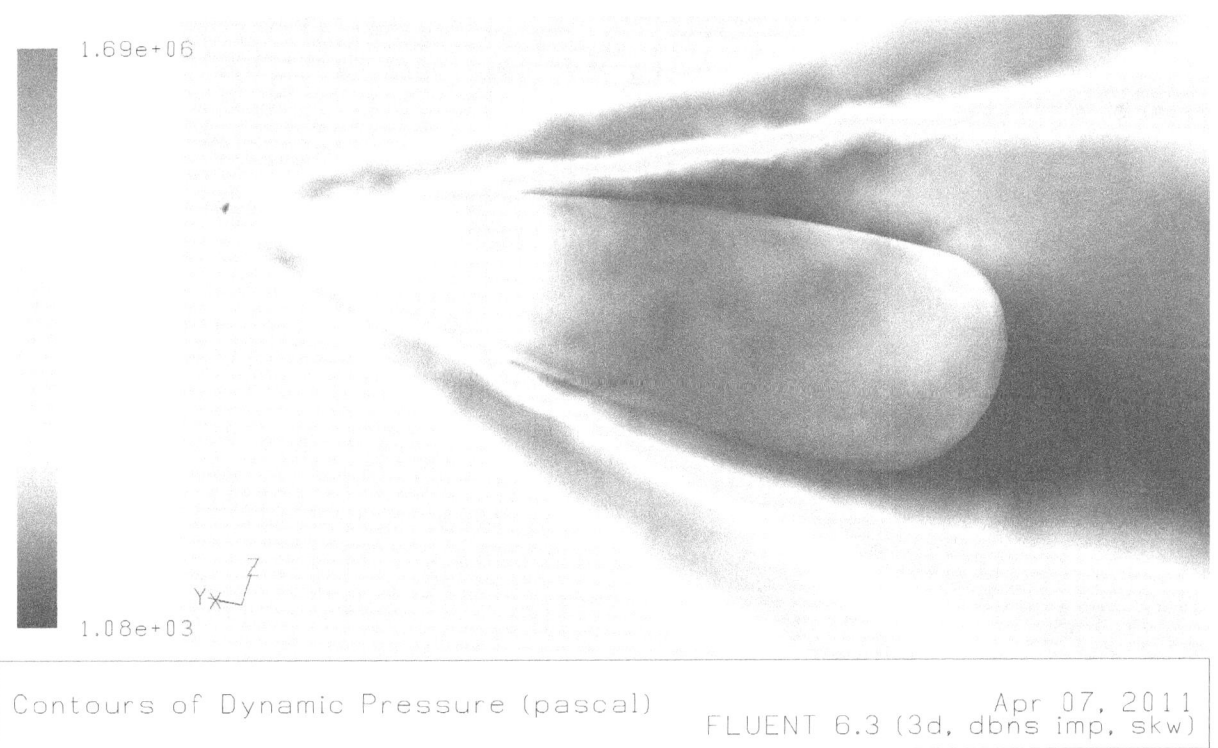

Figure.No.6.4 Contours of Dynamic pressure of semi blunt head bullet at 3.5 Mach and Two degree angle of attack

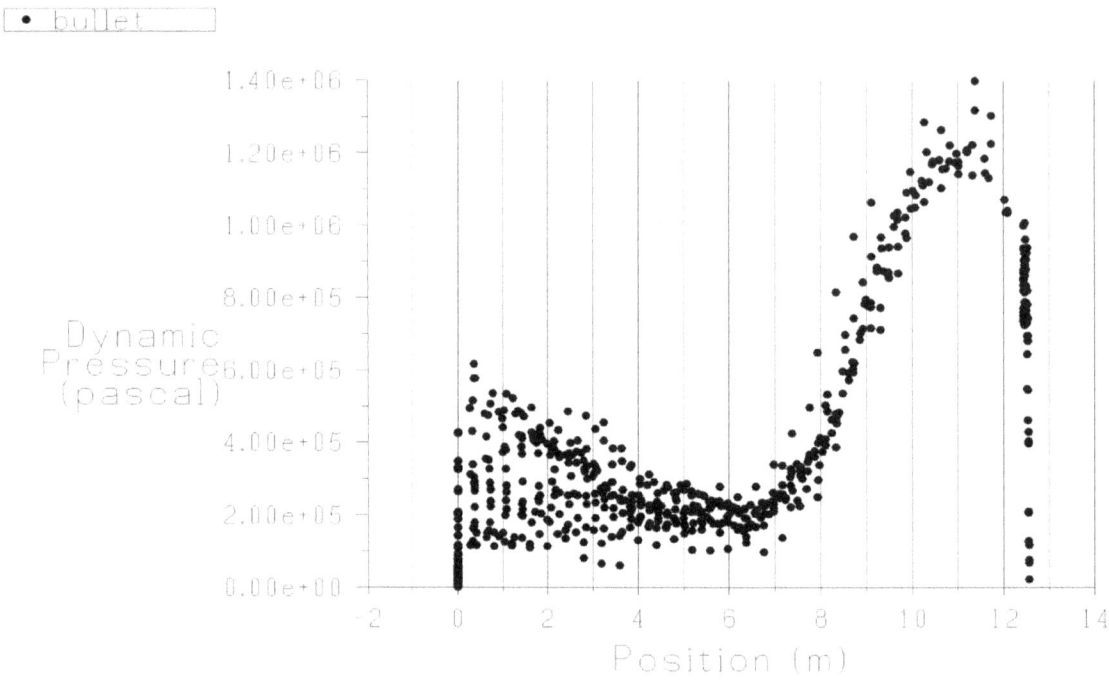

Graph.No.6.4 Dynamic pressure graph

The contour and graph are showing the same phenomena that where dynamic pressure is high static pressure will be low and vice a versa. The maximum dynamic pressure goes till 1.4e+06 pascal.

6.1.1. C 3.5 Mach 2.5 angle of attack

The following figures will show the variation of static and dynamic pressure at 3.5 Mach 2.5-degree angle of attack.

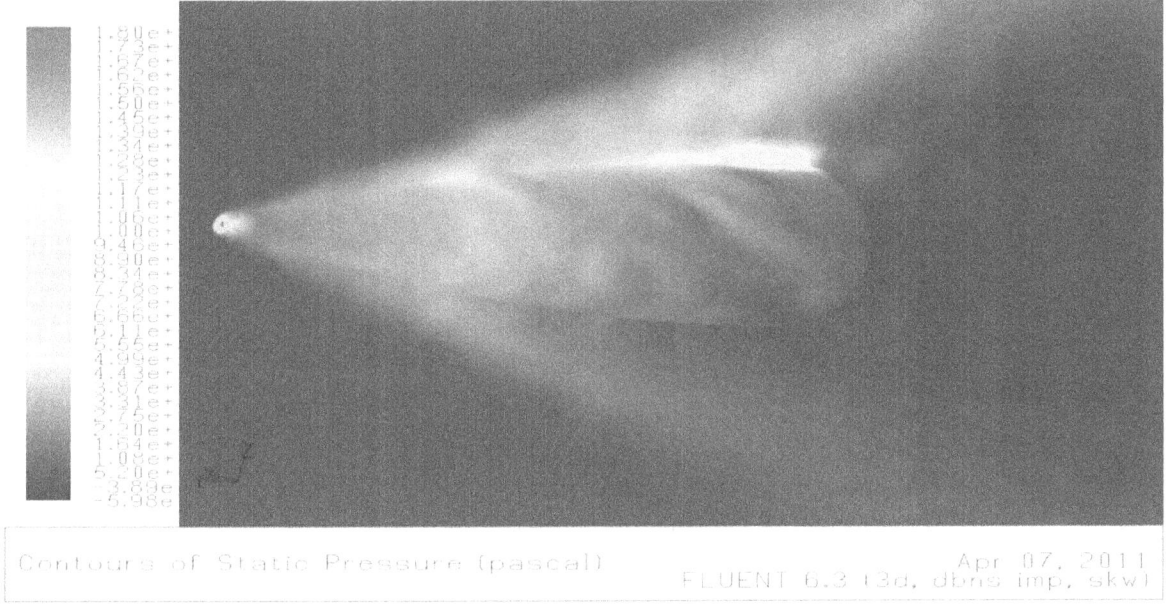

Figure.No.6.5 Contours of static pressure of semi blunt head bullet at 3.5 Mach and 2.5 degree angle of attack

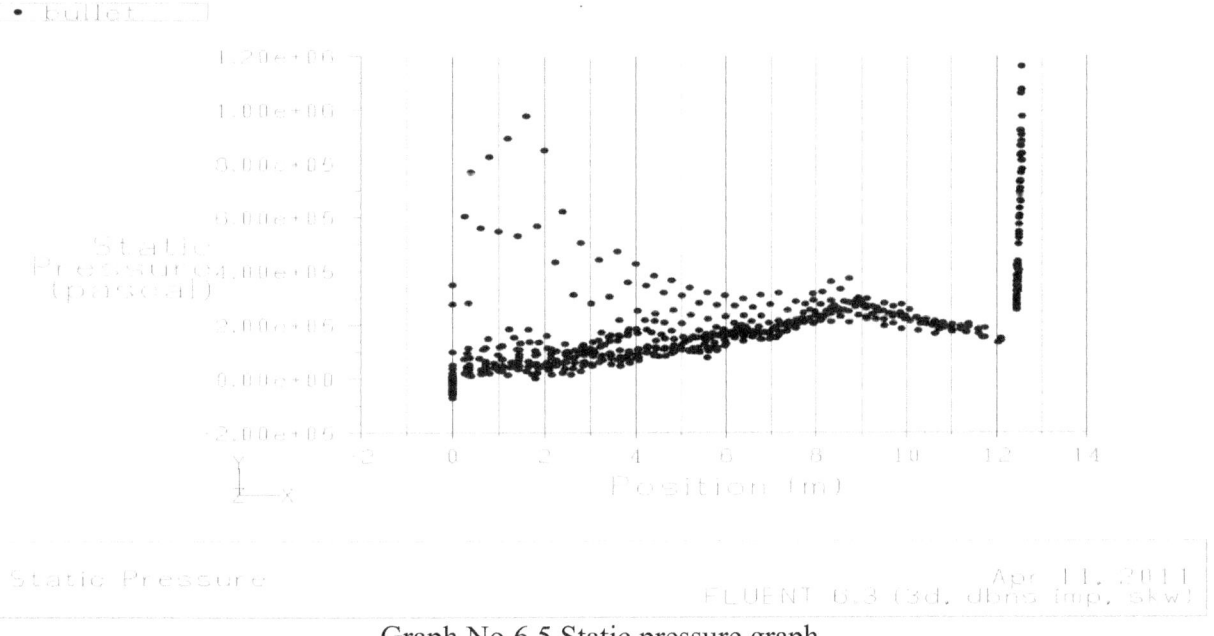

Graph.No.6.5 Static pressure graph

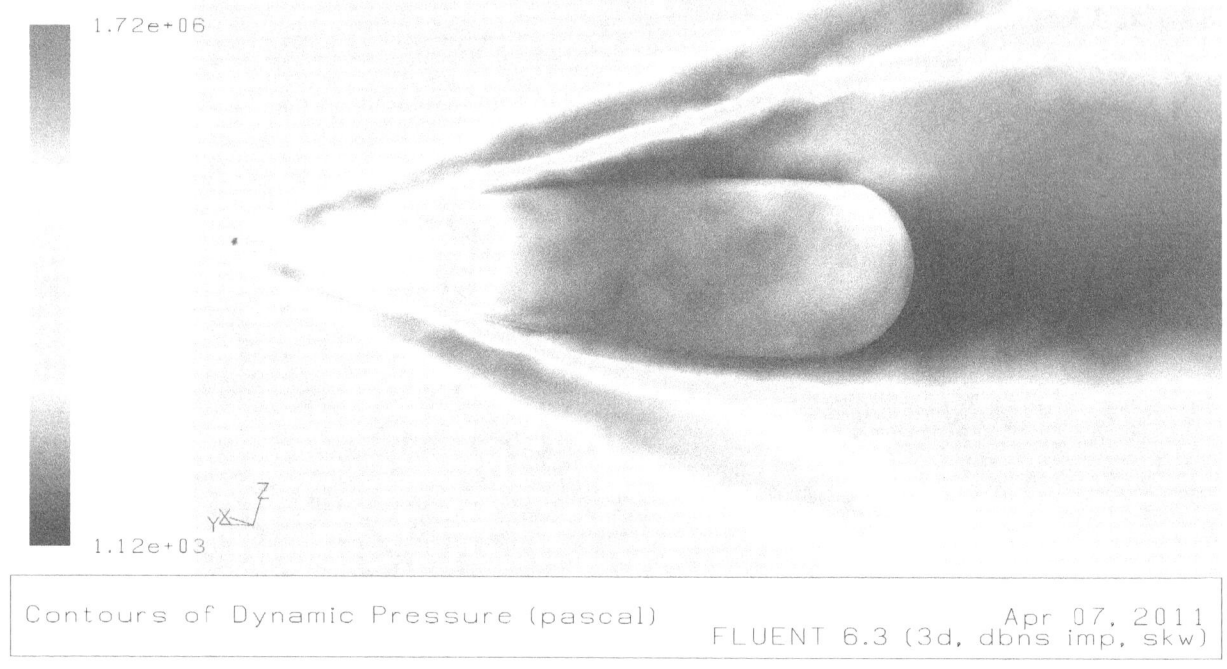

Figure.No.6.6 Contours of dynamic pressure of semi blunt head bullet at 3.5 Mach and 2.5 degree angle of attack

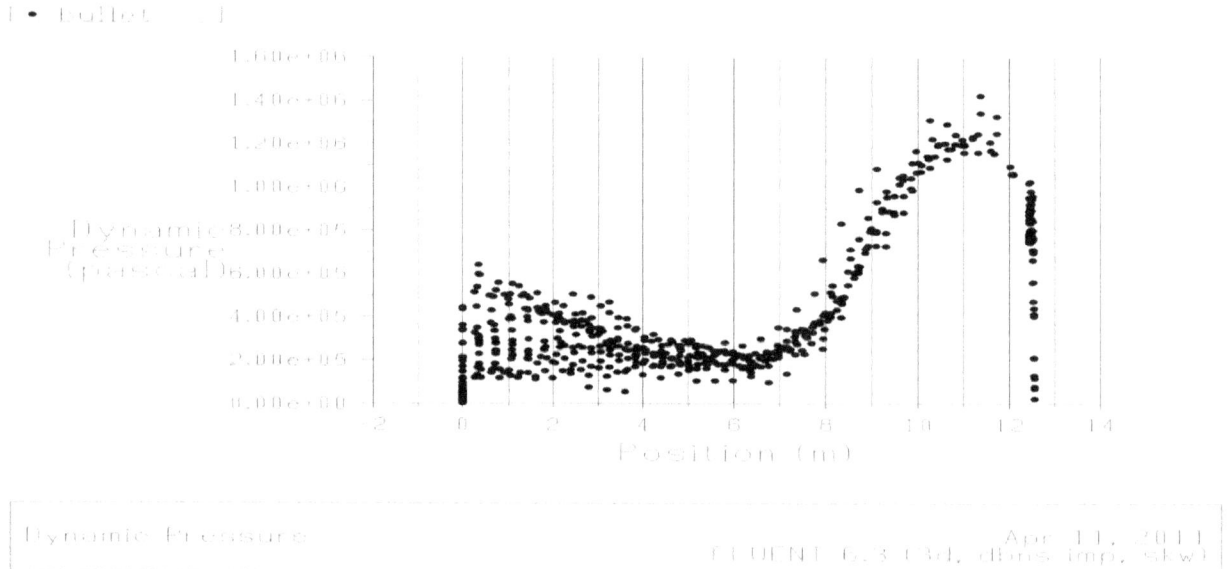

Graph.No.6.6 Dynamic pressure graph

The graph of dynamic pressure shows that the maximum pressure at 2.5-degree angle of attack is 1.35e+06 pascal.

6.1.2.a 3 Mach at 0 angle of attack

The following are showing dynamic and static pressure at 3 Mach at 0 angle of attack.

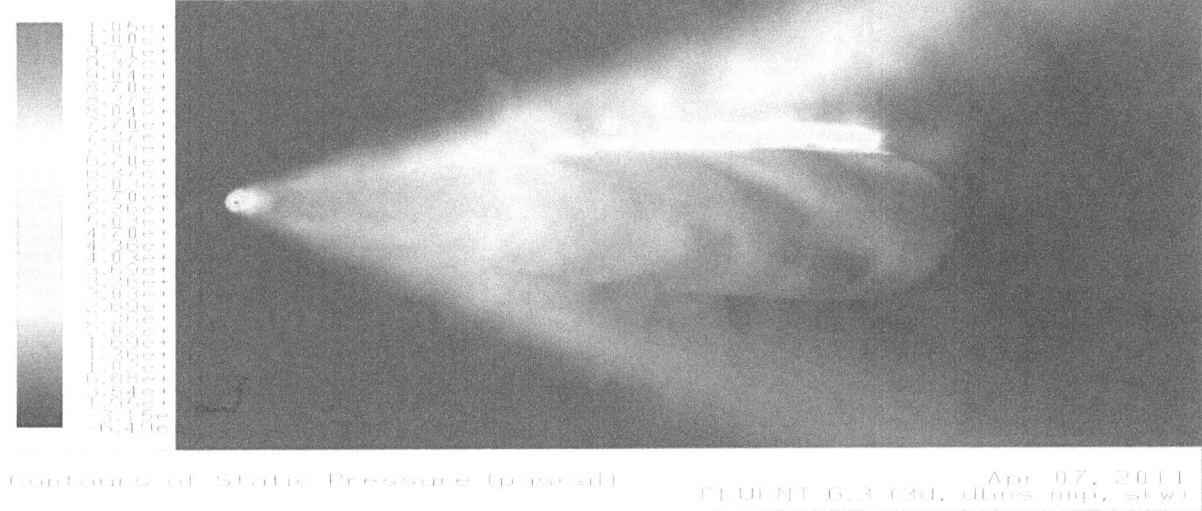

Figure.No.6.7 Contours of static pressure of semi blunt head bullet at 3 Mach and zero angle of attack

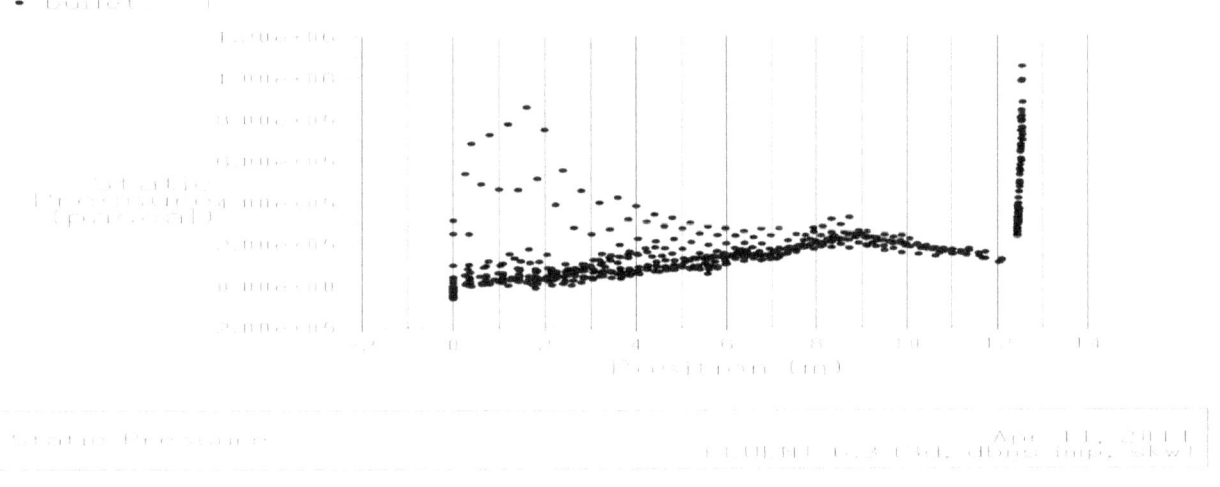

Graph.No.6.7 Static pressure graph

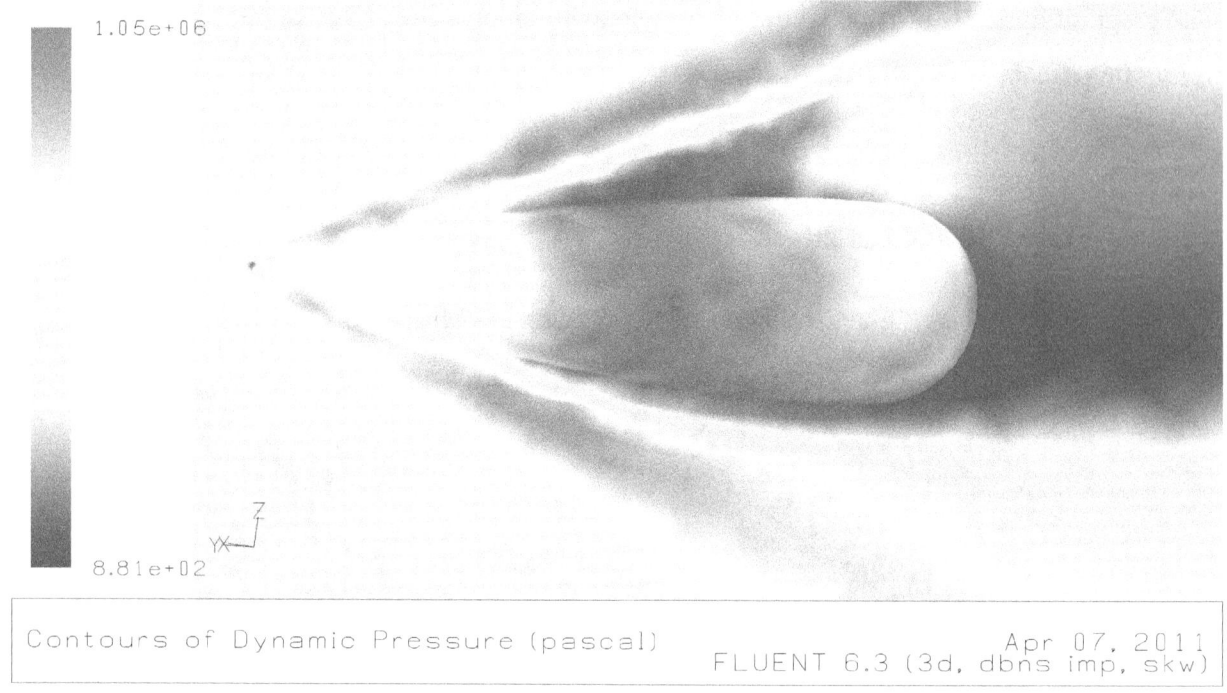

Figure.No.6.8 Contours of dynamic pressure of semi blunt head bullet at 3 Mach and zero angle of attack.

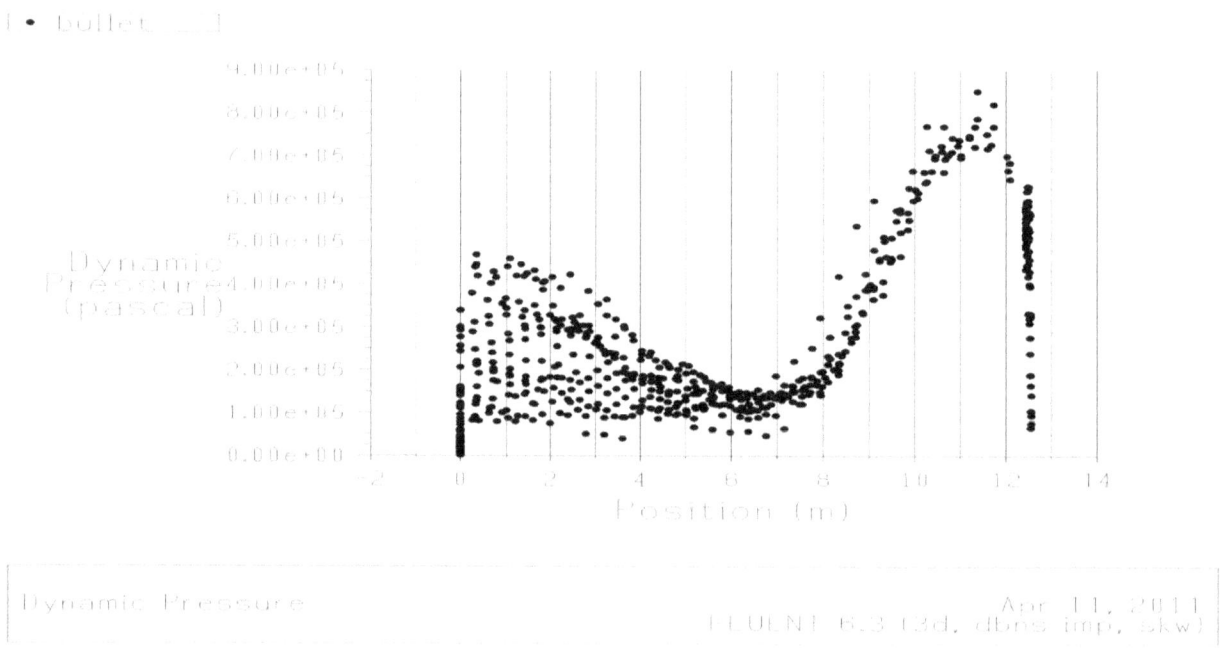

Graph.No.6.8 Dynamic pressure graph

The maximum static pressure is 1.18e+06 pascal and maximum dynamic pressure is 8.9e+05 pascal.

6.1.2.b 3 Mach at 2 degree angle of attack

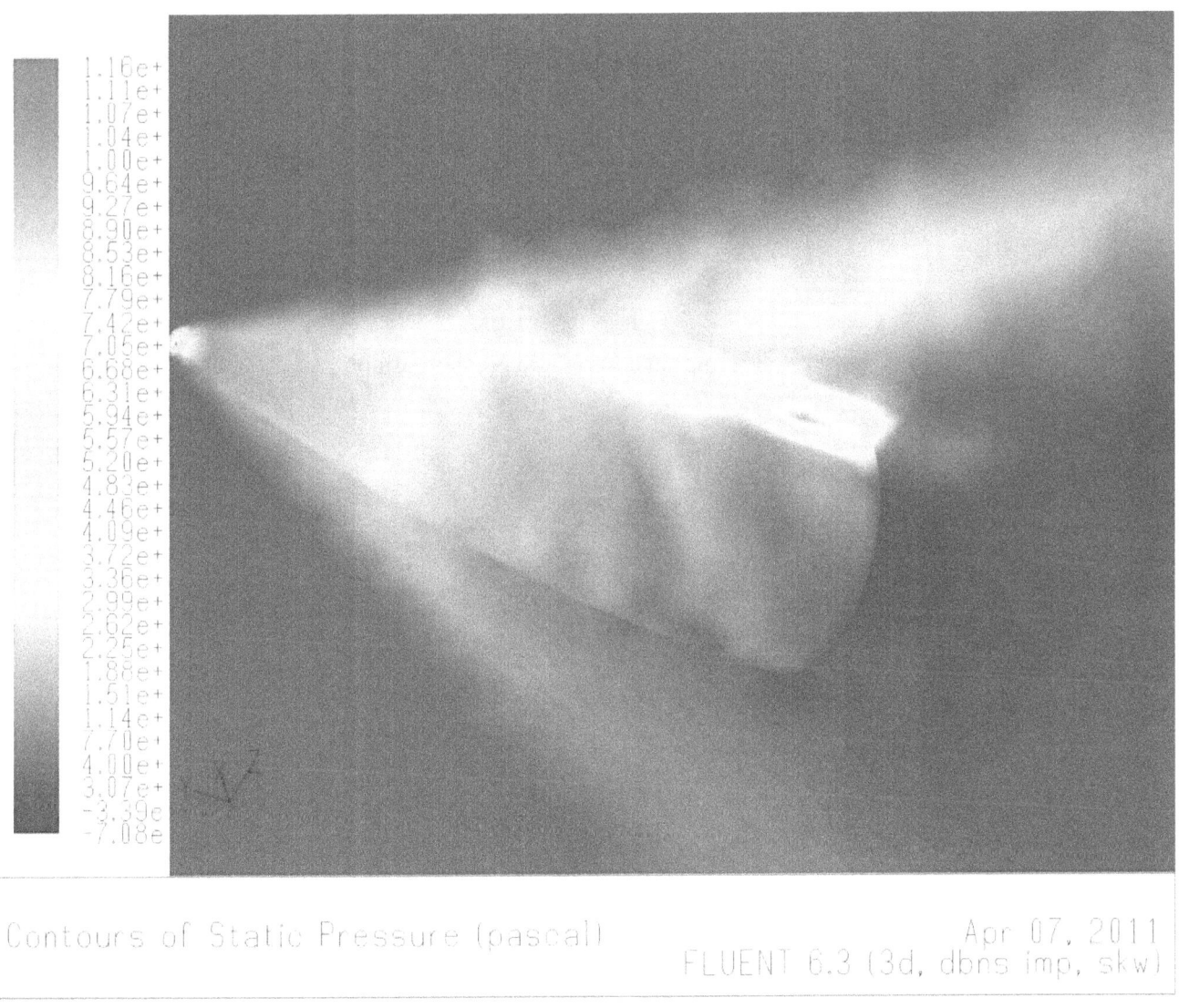

Figure.No.6.9 Contours of static pressure of semi blunt head bullet at 3 Mach and 2 degree angle of attack

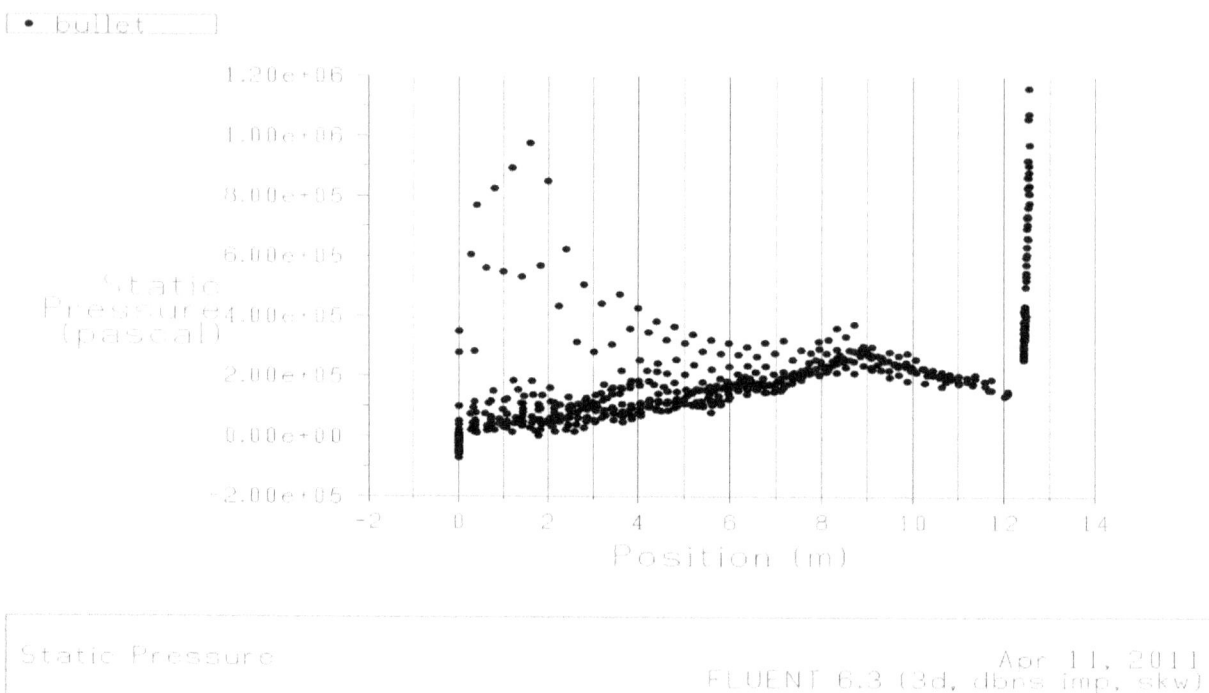

Graph.No.6.9 static pressure graph

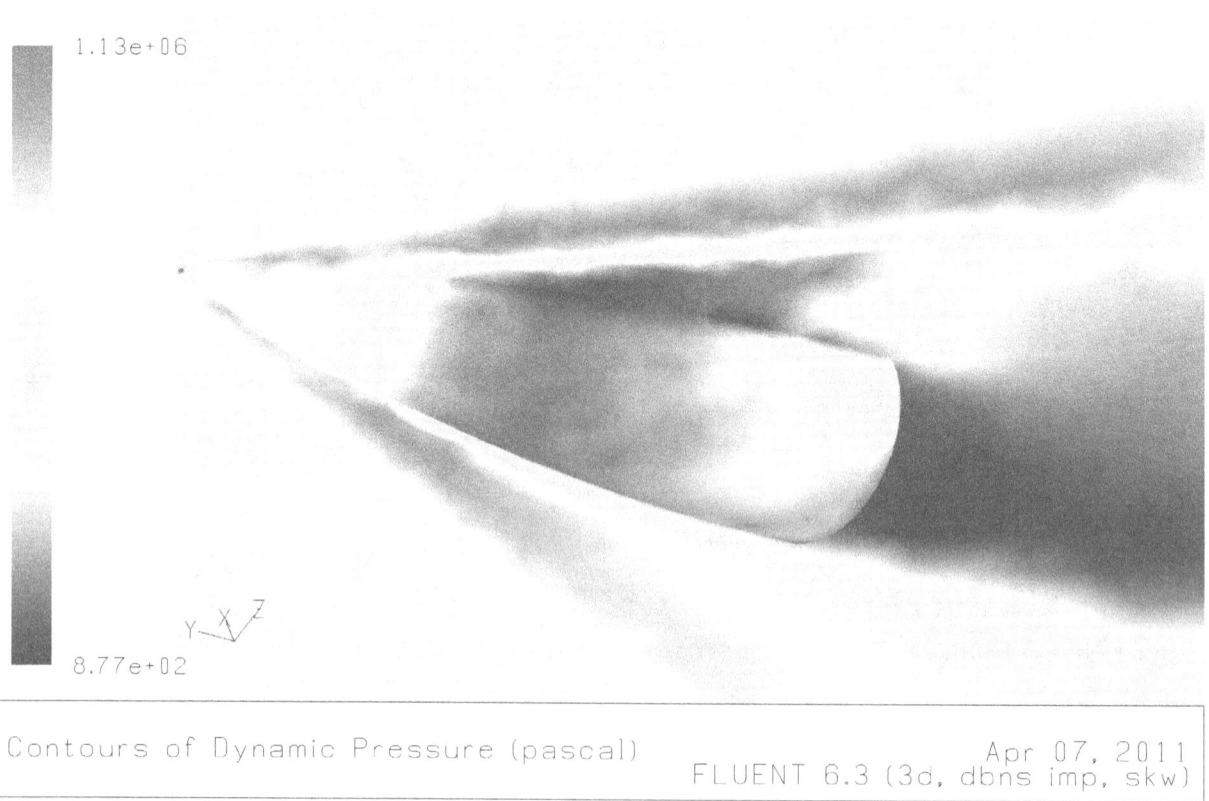

Figure.No.6.10 Contours of dynamic pressure of semi blunt head bullet at 3 mach and 2 degree angle of attack

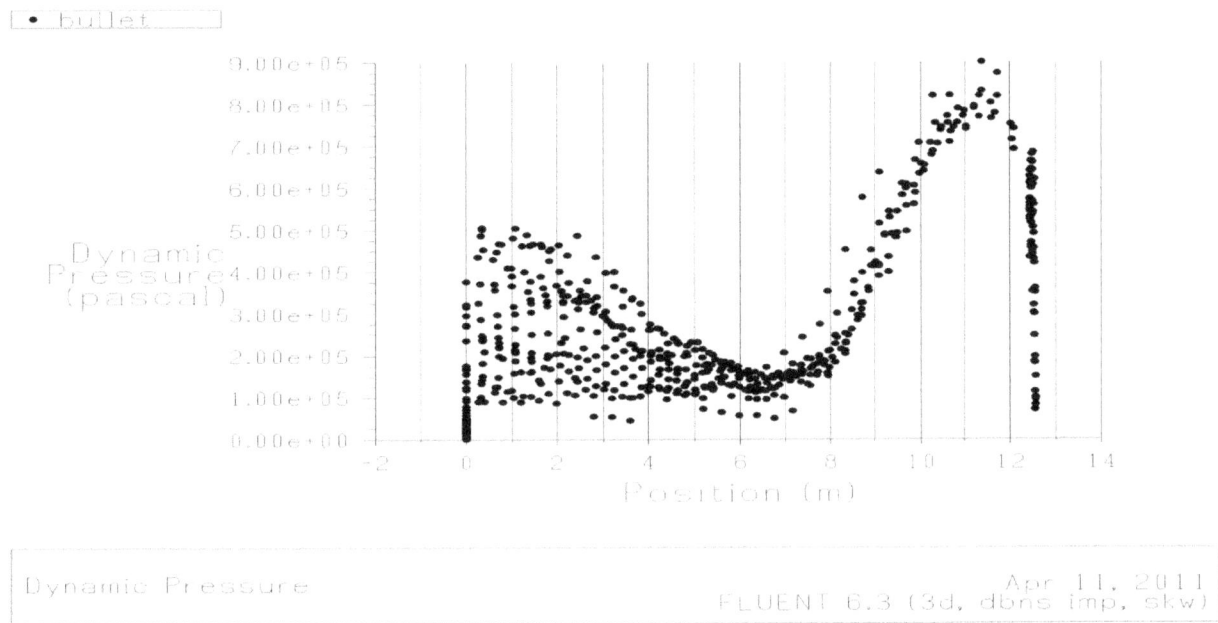

Graph.No.6.10 Dynamic pressure graph

The above figure shows the static pressure and dynamic pressure of projectile (M16 bullet) at 3 Mach at 2 degree angle of attack.

The maximum dynamic pressure on the projectile is 9.00e+05 pascal.

The conclusion derived from above discussion is that the dynamic pressure increases with in Mach number and at a certain Mach no. it will increase with the increase in angle of attack.

CHAPTER 7

Conclusion

7.1 Comparison between shapes

In chapter 5, the point head bullet, pure blunt bullet and semi blunt bullet WERE analyzed.

The contours show that for A long range bullet, neither pure blunt shape nor point head bullet is suitable because in pure blunt shape a strong bow shock is encountered in front of bullet. The properties like pressure, temperature, density and velocity etc. of flow change very dramatically. It will increase the drag which provided hindrance in the motion of projectile. But in pure blunt shape the pressure is distributed over the dome/arc shape. The major advantage of the shape is that the temperature is not concentrated at a point. It is distributed which help to save the integrity of properties of material.

In point head bullet the oblique shocks forms which is weaker than the bow shocks but the formation of boundary layer increase the effective area. The phenomena occur due to displacement thickness of the boundary layer. As the result skin friction drag increases, which is also a curse for long range bullets?

Therefore, a compromise between theM is NEEDED, which suits the requirement of long range bullets with best penetration effect. For that semi blunt shape is analyzed, which had advantages of both.

7.2 Analysis of Semi blunt shape projectile at different Mach No and at different angle of attack

In chapter 6, THE pressure change of the semi blunt shape is analyzed. First the projectile is analyzed at 3.5 Mach at 0 degree angle of attack. The static pressure is maximum at stagnation point. It is 1.18e+06 pascal at stagnation point which is at 12.7 mm from the origin (length of bullet is 12.7mm). There is sudden drop in static pressure at 12mm and the value comes down to 2.00e+05 pascal. It is because at the stagnation point velocity is zero and therefore pressure is at its maximum value.

Than flow is analyzed at 2 degree angle of attack. The contour and graph is displayed the phenomena that where dynamic pressure is high static pressure will be low and vice a versa. The maximum dynamic pressure goes till 1.4e+06 pascal. At 2.5 degree angle of attack maximum dynamic pressure is 1.35e+06 pascal.

At 3 Mach flow was also analyzed, the phenomena came into focus into is that "with increase of Mach number the pressure increases, and if Mach number is kept constant than pressure increases with the angle of attack.

7.3 The Comparison of the dropping of bullet

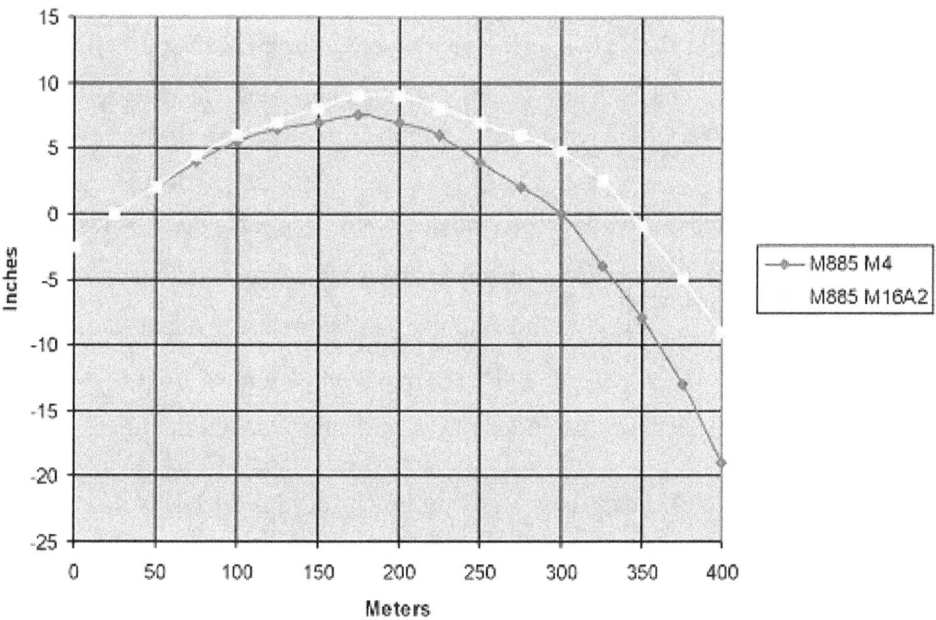

Figure 7.1: The drop of M16 projectile; pre measured

The above figure shows the drop of projectile fired from M 16A2 rifle and M4. The drop of bullet fired from M16A2 drop of 7.5 inch in 400 metres and the bullet fired from M4 had drop of 15 inches. These results are pre-established and the figure below shows the drop measured by the program of trajectory analysis (6 degree of freedom model).

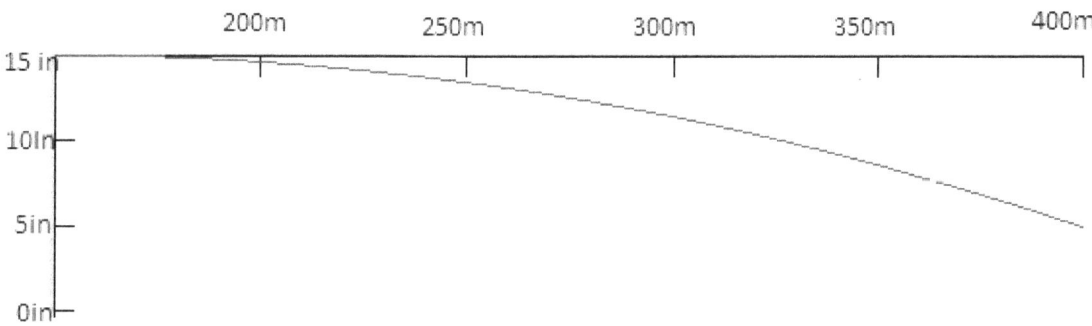

Figure 7.2: The drop of M16 bullet measured by MATLAB program

The drop measured by MATLAB is 8 inchs at 400m. Error percentage is 6.667 %.

This shows that condition taken and modeling has 6.67% error.

FUTURE SCOPE

In conclusion, the application of Computational Fluid Dynamics (CFD) analysis to the study of bullets at high speeds has proven to be an invaluable tool for understanding and optimizing their aerodynamic performance. The complex fluid flow phenomena encountered during high-speed flight pose significant challenges, but CFD offers a comprehensive and efficient approach to analyze and improve bullet designs.

Through CFD analysis, researchers and engineers have gained deep insights into the aerodynamic forces acting on bullets as they travel at supersonic velocities. By accurately modeling and simulating the flow around the bullet, factors such as drag, lift, and stability can be precisely quantified. This understanding has paved the way for the development of bullet designs that minimize drag and maximize stability, ultimately leading to improved accuracy, extended range, and enhanced terminal ballistics.

One of the key advantages of CFD analysis in studying bullets at high speeds is its ability to capture and analyze the shockwaves and vortices generated by the bullet's flight. These complex flow features play a crucial role in determining the bullet's stability, particularly as it transitions from subsonic to supersonic speeds. By accurately simulating these flow phenomena, engineers can optimize bullet shapes, such as ogive profiles and boat-tail designs, to mitigate shockwave interference and reduce destabilizing effects.

Furthermore, CFD analysis allows for the evaluation and optimization of bullet designs under varying flight conditions. The ability to predict and assess the effects of variables such as altitude, temperature, humidity, and wind on bullet trajectories is invaluable in military operations, where accurate long-range shooting is often required in diverse and challenging environments. By considering these factors in the design process, engineers can develop bullets that exhibit robust performance across a wide range of operational conditions.

Another significant benefit of CFD analysis in the study of high-speed bullets is its role in reducing undesirable effects such as sonic boom, muzzle blast, and recoil. By accurately simulating and analyzing the flow dynamics involved in these phenomena, engineers can design muzzle devices and bullet configurations that minimize noise and recoil, thereby enhancing shooter comfort and reducing the environmental impact.

Moreover, CFD analysis enables cost-effective and efficient exploration of design alternatives. The ability to virtually test and iterate through different bullet designs at high speeds significantly reduces the time and resources required for physical testing. This accelerates the development process and facilitates innovation, enabling engineers to explore a wider design space and uncover novel solutions.

As CFD methodologies and computational resources continue to advance, the accuracy and fidelity of bullet analysis at high speeds will further improve. The integration of multiphysics simulations, such as fluid-structure interaction, will provide a more comprehensive understanding of bullet behavior and enable the optimization of bullet materials and structural integrity.

In conclusion, CFD analysis of bullets at high speeds has become an essential tool in the field of ballistics. Its ability to accurately model and simulate the complex fluid flow phenomena encountered during high-speed flight has revolutionized bullet design and performance optimization. The insights gained from CFD analysis have led to improved accuracy, extended range, reduced undesirable effects, and enhanced bullet stability under various environmental conditions. As the field of CFD continues to evolve, it holds tremendous potential for further advancements in bullet technology, ultimately contributing to more effective and efficient firearms systems.

Appendix A

A.1 Program of 6 degree of freedom for projectile.

```
clear all
close all
clc

%%% MISSILE PARAMETERS %%%

d=0.00566;
area=pi*d*d/4;
mass=.01;
Ix=1.62e-08;
Iy=5.9e-09;
Iz=1.62e-08;
Ixz=0;

%%% INITIAL MISSILE CONDITIONS %%%

x=0;
y=0;
z=10;
u=900;
v=0;
w=0.0;
thrustx=0;
thrusty=0;
thrustz=0;
p=0;
q=0.000;
r=0.000;
theta=0;
phi=0;
psi=0;
cX=0;
cY=0;
cZ=0;
```

```
cL=0;
cM=0.0;
cN=0.00;

%%% GENERAL %%%

g=9.80665;
j=1;
rho0=1.752;
timestep=0.01;

cph = cos(phi);
sph = sin(phi);
cth = cos(theta);
sth = sin(theta);
cps = cos(psi);
sps = sin(psi);
dcm = [cth*cps               cth*sps              -sth
       sph*sth*cps-cph*sps   sph*sth*sps+cph*cps   sph*cth
       cph*sth*cps+sph*sps   cph*sth*sps-sph*cps   cph*cth];
dcm=dcm';
% alp0=acos(u/sqrt(u^2+v^2+w^2));
% alp=0;
% if(v==0)
%         ph0=0;
%     elseif(w==0)
%         ph0=pi/2;
%     else
%         ph0=atan(v/w);
%     end
% ph=0;

%%% PATH CALCULATION %%%

while(z>0)

    if(z<11000) ;
```

```
        density=1.225*(1.-.000022557695*z)^(4.25587);
    else
        density=.36392*exp(-1.576584E-04*(z-11000));
    end;


    %density=rho0*exp(((-z)/6700));

    alpha=acos(u/sqrt(u^2+v^2+w^2));
%     alp=alpha-alp0;
%     alp0=alpha;
    if(v==0)
        PHI_=0;
    elseif(w==0)
        PHI_=pi/2;
    else
        PHI_=atan(v/w);     %body axis bank angle
    end
%     ph=PHI_-ph0;
%     ph0=PHI_;
%
    if(z<=11000)
        lapse=6.5;
    end
    if(z>11000)
        lapse=0;
    end
    temp=288.2-(lapse*z/1000);
    vs=sqrt(1.4*287*temp);
    mach=(sqrt((u^2)+(v^2)+(w^2)))/vs;


    velocity=(sqrt((u^2)+(v^2)+(w^2)));    % Resultant velocity

    % Forces and Moments in Inertial Frame
    %alp*180/2;
```

```
[cX,cY,cZ,cL,cM,cN]=interpolation(sqrt((alpha*180/pi)*(alpha*180/pi)),mach,0)
; % Calling function for interpolation
    cM=cM/1000;
    %velocity=(sqrt((u^2)+(w^2)));      % Resultant velocity


    X=0.5*cX*density*velocity*velocity*area;
    Y=0.5*cY*density*velocity*velocity*area;
    Z=0.5*cZ*density*velocity*velocity*area;


    L=0.5*cL*density*velocity*velocity*area*d/2;
    M=0.5*cM*density*velocity*velocity*area*d/2;
    N=0.5*cN*density*velocity*velocity*area*d/2;



    if(u==0)
        alp=0;
    elseif(w==0)
        alp=pi/2;
    else
        alp=atan(u/w);
    end



% % %     X=X1*sin(alp)-Y1*cos(PHI_)*cos(alp)-Z1*sin(PHI_)*cos(alp);
% % %     Y=X1*cos(alp)-Y1*cos(PHI_)*sin(alp)-Z1*sin(PHI_)*sin(alp);
% % %     Z=-Y1*sin(PHI_)+Z1*cos(PHI_);
% % %     L=L1*sin(alp)-M1*cos(PHI_)*cos(alp)-N1*sin(PHI_)*cos(alp);
% % %     M=L1*cos(alp)-M1*cos(PHI_)*sin(alp)-N1*sin(PHI_)*sin(alp);
% % %     N=-M1*sin(PHI_)+N1*cos(PHI_);
% % %

     sal=sin(alp);
    cal=cos(alp);
    spi=sin(PHI_);
    cpi=cos(PHI_);
```

```
T=[cal   sal*spi   -sal*cpi
   0     cpi        spi
   sal   -cal*spi   cal*cpi];

E=T*[X
     Y
     Z];
X=E(1);
Y=E(2);
Z=E(3);


E=T*[L
     M
     N];

L=E(1);
M=E(2);
N=E(3);


E=dcm*[X
       Y
       Z];
X=E(1);
Y=E(2);
Z=E(3);


E=dcm*[L
       M
       N];

L=E(1);
M=E(2);
N=E(3);
```

```
%
%      %%% Rigid body equations %%%

I=[Ix 0 -Ixz
   0 Iy 0
   -Ixz 0 Iz];


A=[(L-(q*r*(Iz-Iy))+(Ixz*p*q))
   (M-(r*q*(Ix-Iz))-(Ixz*((p^2)-(r^2))))
   (N-(p*q*(Iy-Ix))-(Ixz*q*r))];


wx=I\A;


prate=wx(1);
qrate=wx(2);
rrate=wx(3);


p1=p+(timestep*prate);
q1=q+(timestep*qrate);
r1=r+(timestep*rrate);


cph = cos(phi);
sph = sin(phi);
cth = cos(theta);
sth = sin(theta);
cps = cos(psi);
sps = sin(psi);



thetarate=(q*cph)-(r*sph);
phirate=p+(q*sph*sth/cth)+(r*cph*sth/cth);
psirate=((q*sph)+(r*cph))/cth;


theta1=theta+(thetarate*timestep);
phi1=phi+(phirate*timestep);
psi1=psi+(psirate*timestep);
```

```
Th=[thrustx
    thrusty
    thrustz];
The=dcm*Th;
Tx=The(1);
Ty=The(2);
Tz=The(3);


urate=((Tx+X-(mass*g*sth))/mass)-(q*w)+(r*v);
vrate=((Ty+Y+(mass*g*cth*sph))/mass)-(r*u)+(p*w);
wrate=((Ty+Z+(mass*g*cth*cph))/mass)-(p*v)+(q*u);


u1=u+(urate*timestep);
v1=v+(vrate*timestep);
w1=w+(wrate*timestep);



dcm = [cth*cps            cth*sps              -sth
    sph*sth*cps-cph*sps  sph*sth*sps+cph*cps  sph*cth
    cph*sth*cps+sph*sps  cph*sth*sps-sph*cps  cph*cth];

dcm=dcm';

V=[u1
   v1
   w1];

wx=dcm*V;

xrate=wx(1);
yrate=wx(2);
zrate=wx(3);


x1=x+(xrate*timestep);
```

```
        y1=y+(yrate*timestep);
        z1=z-(zrate*timestep);



        time(j)=(j-1)*(timestep);

        xm(j)=x;
        ym(j)=y;
        zm(j)=z;
        phim(j)=phi;
        psim(j)=psi;
        thetam(j)=theta;


        j=j+1;

        %%% Assigning new values to all the variables %%%
        p=p1;
        q=q1;
        r=r1;
        theta=theta1;
        phi=phi1;
        psi=psi1;
        x=x1;
        y=y1;
        z=z1;
        u=u1;
        v=v1;
        w=w1;

end


%%% PLOTS %%%
plot3 (xm,ym,zm);
figure
plot(time,zm);xlabel('time');ylabel('z');
figure
plot(time,ym);xlabel('time');ylabel('y');
```

```
figure
plot(time,xm);xlabel('time');ylabel('x');
% figure
% subplot(2,2,1), plot(time,phim);xlabel('time');ylabel('PHI');
% subplot(2,2,2), plot(time,psim);xlabel('time');ylabel('PSI');
% subplot(2,2,3), plot(time,thetam);xlabel('time');ylabel('THETA');
```

A.2 Interpolation of forces

```
function [cX,cY,cZ,cL,cM,cN]=interpolation(alpha,mach)

    if(degree==0)
        if(mach>=1.33 && mach<3)
            x1=1.33;
            x2=3;
            r1=1;
            r2=2;
        end
        if(mach<1.33)
            x1=1.33;
            x2=3;
            r1=1;
            r2=2;
        end

        if(mach>=3)
            x1=3;
            x2=3.5;
            r1=2;
            r2=3;
        end
        if(alpha<0)
            y1=0;
            y2=1.33;
            c1=1;
            c2=2;
```

```
end

if(alpha>=0 && alpha<1.33)
    y1=0;
    y2=1.33;
    c1=1;
    c2=2;
end

if(alpha>=1.33 && alpha<2)
    y1=1.33;
    y2=2;
    c1=2;
    c2=3;
end

if(alpha>=2 )
    y1=2;
    y2=2.5;
    c1=3;
    c2=4;
end



cmx=[0.00204 -0.00076 0.00336 0.00136
    0.00000 0.00092 0.00101 0.00189
    0.00236 0.00109 0.00142 -0.00086
     ];

cmy=[0.01015 -0.44229 -1.15826 -1.9417
    0.01845 -0.45197 -1.24018 -2.1855
    -0.00002 -0.46472 -1.33761 -2.27
     ];

cmz=[-0.01418 0.01658 0.03876 0.01505
    0.00112 0.03574 0.01096 0.02770
    -0.00109 0.00136 0.01640 0.02405
```

```
        ];

    cx=[-0.09546 -0.09774 -0.09554 -0.0887
        -0.08862 -0.09402 -0.09374 -0.08921
        -0.10257 -0.11091 -0.10550 -0.10167
        ];

    cy=[0.002943 -0.002787 -0.007151 -0.00316
        0.000010 -0.001720 -0.004223 -0.009807
        0.000000 -0.000305 -0.002473 -0.007891
        ];

    cz=[0.002285 -0.271823 -0.592251 -0.947
        -0.000012 -0.271036 -0.602926 -0.998
        0.000003 -0.268833 -0.627922 -1.022113
        ];
end




    if(mach>3.5)
        mach=3.5;
    end
    if(alpha>2.5)
        alpha=2.5;
    end
    if(mach<1.33)
        mach=1.33;
    end
    if(alpha<0)
        alpha=0;
    end
    q1=((cx(r1,c1)*(mach-x1))+(cx(r2,c1)*(x2-mach)))/(x2-x1);
    q2=((cx(r1,c2)*(mach-x1))+(cx(r2,c2)*(x2-mach)))/(x2-x1);
```

```
cX=((q1*(alpha-y1))+(q2*(y2-alpha)))/(y2-y1);


q1=((cy(r1,c1)*(mach-x1))+(cy(r2,c1)*(x2-mach)))/(x2-x1);
q2=((cy(r1,c2)*(mach-x1))+(cy(r2,c2)*(x2-mach)))/(x2-x1);
cY=((q1*(alpha-y1))+(q2*(y2-alpha)))/(y2-y1);


q1=((cz(r1,c1)*(mach-x1))+(cz(r2,c1)*(x2-mach)))/(x2-x1);
q2=((cz(r1,c2)*(mach-x1))+(cz(r2,c2)*(x2-mach)))/(x2-x1);
cZ=((q1*(alpha-y1))+(q2*(y2-alpha)))/(y2-y1);


q1=((cmx(r1,c1)*(mach-x1))+(cmx(r2,c1)*(x2-mach)))/(x2-x1);
q2=((cmx(r1,c2)*(mach-x1))+(cmx(r2,c2)*(x2-mach)))/(x2-x1);
cL=((q1*(alpha-y1))+(q2*(y2-alpha)))/(y2-y1);


q1=((cmy(r1,c1)*(mach-x1))+(cmy(r2,c1)*(x2-mach)))/(x2-x1);
q2=((cmy(r1,c2)*(mach-x1))+(cmy(r2,c2)*(x2-mach)))/(x2-x1);
cM=((q1*(alpha-y1))+(q2*(y2-alpha)))/(y2-y1);


q1=((cmz(r1,c1)*(mach-x1))+(cmz(r2,c1)*(x2-mach)))/(x2-x1);
q2=((cmz(r1,c2)*(mach-x1))+(cmz(r2,c2)*(x2-mach)))/(x2-x1);
cN=((q1*(alpha-y1))+(q2*(y2-alpha)))/(y2-y1);


end
```

APPENDIX B: The Values for Aerodynamic Analysis of Bullets

In the analysis of bullets' aerodynamic performance, several key parameters and values play a crucial role. The following list provides an overview of the common values used in aerodynamic analysis of bullets:

1. **Coefficient of Drag (Cd):** The coefficient of drag represents the resistance encountered by a bullet as it moves through the air. It quantifies the aerodynamic drag force acting on the bullet and is a crucial parameter in bullet design optimization. The value of Cd varies with bullet shape, velocity, and Reynolds number.

2. **Ballistic Coefficient (BC):** The ballistic coefficient is a measure of a bullet's ability to overcome air resistance. It is derived from the Cd and the bullet's mass, diameter, and sectional density. The BC is a key parameter for predicting bullet trajectory, range, and wind drift.

3. **Lift Coefficient (Cl):** The lift coefficient quantifies the lift force generated on a bullet due to its spin. The spin imparts gyroscopic stability to the bullet, reducing its tendency to tumble or deviate from its intended path. The value of Cl depends on the bullet's rotational velocity and spin rate.

4. **Magnus Effect**: The Magnus effect refers to the deviation in the trajectory of a spinning bullet caused by the pressure difference on its upper and lower surfaces. It leads to bullet drift, especially at longer ranges. The Magnus effect can be quantified using the Magnus coefficient, which depends on the bullet's spin rate, diameter, and velocity.

5. **Transitional Flow**: The transition from subsonic to supersonic flow is a critical phase in bullet flight. The critical Mach number (Mcr) represents the speed at which a bullet transitions from subsonic to supersonic flow. Understanding and predicting this transition is essential for accurate aerodynamic analysis.

6. **Reynolds Number (Re):** The Reynolds number characterizes the flow regime around a bullet and helps determine whether the flow is laminar or turbulent. It is calculated using the bullet's diameter, velocity, and kinematic viscosity of the air. The value of Re affects the accuracy of turbulence modeling in CFD simulations.

7. **Mach Number (M):** The Mach number represents the ratio of a bullet's velocity to the speed of sound in the surrounding air. The Mach number determines whether the bullet is traveling at subsonic, transonic, or supersonic speeds, influencing the flow characteristics and aerodynamic forces acting on the bullet.

8. **Bullet Shape:** The geometry and profile of a bullet have a significant impact on its aerodynamic performance. Parameters such as ogive shape, boat-tail angle, meplat diameter, and base design affect the flow separation, drag, and stability of the bullet.

It is important to note that the values mentioned above are not fixed and can vary depending on specific bullet designs, calibers, and intended applications. Experimental data, wind tunnel testing, and CFD simulations are often used to determine and refine these values for a given bullet design.

The precise determination of these parameters is essential for accurate CFD analysis of bullet aerodynamics, allowing for more reliable predictions of bullet behavior, trajectory, stability, and performance in various shooting scenarios.

Disclaimer: The values provided in this appendix are general in nature and may not be applicable to specific bullet designs or proprietary data. It is recommended to consult relevant literature, bullet manufacturers, or conduct detailed testing and analysis for specific bullet aerodynamic characteristics.

References

K. G. Guderley,1950,"Axial symmetric flow patterns at a free stream mach no ,1"USAF. Tech.Report6285

I.M.Ryzhik, 1988"Three dimensional Transonic flow over slender wings and bodies"NASA.Tech.Rept.13118

H. Lomax, 1990, "Linearized compressible flow theory for sonic speed" NASA . REP 956

Url 1:< http// www.wikipedia.com/jpeg.pic>

Url 2: <http// www.iaa.org>

Url 3:<http://www.virtualdub.org >

Ur l4: <http://www.edwdebono.com/debono/selfi.htm>

Url 5: <http// www.pbs.org/wgbh/nova/barrier.html>

www.ingramcontent.com/pod-product-compliance
Lightning Source LLC
Chambersburg PA
CBHW062114220526
45471CB00010B/3737